T0341327

Machine Learning and Deep Learning Techniques in Wireless and Mobile Networking Systems

Big Data for Industry 4.0: Challenges and Applications

Series Editors: Sandhya Makkar, K. Martin Sagayam, and Rohail Hassan

Industry 4.0 or fourth industrial revolution refers to interconnectivity, automation and real time data exchange between machines and processes. There is a tremendous growth in big data from internet of things (IoT) and information services which drives the industry to develop new models and distributed tools to handle big data. Cutting-edge digital technologies are being harnessed to optimize and automate production including upstream supply-chain processes, warehouse management systems, automated guided vehicles, drones etc. The ultimate goal of industry 4.0 is to drive manufacturing or services in a progressive way to be faster, effective and efficient that can only be achieved by embedding modern day technology in machines, components, and parts that will transmit real-time data to networked IT systems. These, in turn, apply advanced soft computing paradigms such as machine learning algorithms to run the process automatically without any manual operations.

The new book series will provide readers with an overview of the state-of-the-art in the field of Industry 4.0 and related research advancements. The respective books will identify and discuss new dimensions of both risk factors and success factors, along with performance metrics that can be employed in future research work. The series will also discuss a number of real-time issues, problems and applications with corresponding solutions and suggestions. Sharing new theoretical findings, tools and techniques for Industry 4.0, and covering both theoretical and application-oriented approaches. The book series will offer a valuable asset for newcomers to the field and practicing professionals alike. The focus is to collate the recent advances in the field, so that undergraduate and postgraduate students, researchers, academicians, and Industry people can easily understand the implications and applications of the field.

Industry 4.0 Interoperability, Analytics, Security, and Case Studies
Edited by G. Rajesh, X. Mercilin Raajini, and Hien Dang

Big Data and Artificial Intelligence for Healthcare Applications
Edited by Ankur Saxena, Nicolas Brault, and Shazia Rashid

Machine Learning and Deep Learning Techniques in Wireless and Mobile Networking Systems
Edited by K. Suganthi, R. Karthik, G. Rajesh, and Peter Ho Chiung Ching

For more information on this series, please visit: www.routledge.com/Big-Data-for-Industry-4.0-Challenges-and-Applications/book-series/CRCBDICA

Machine Learning and Deep Learning Techniques in Wireless and Mobile Networking Systems

Edited by
K. Suganthi, R. Karthik, G. Rajesh,
and Peter Ho Chiung Ching

CRC Press
Taylor & Francis Group
Boca Raton London New York

CRC Press is an imprint of the
Taylor & Francis Group, an **informa** business

First edition published 2022
by CRC Press
6000 Broken Sound Parkway NW, Suite 300, Boca Raton, FL 33487–2742

and by CRC Press
2 Park Square, Milton Park, Abingdon, Oxon, OX14 4RN

© 2022 selection and editorial matter, K. Suganthi, R. Karthik, G. Rajesh, and Ho Chiung Ching; individual chapters, the contributors

CRC Press is an imprint of Taylor & Francis Group, LLC

Library of Congress Cataloging-in-Publication Data
Names: Suganthi, K., editor.
Title: Machine learning and deep learning techniques in wireless and mobile
   networking systems / edited by K. Suganthi, R. Karthik, G. Rajesh, and
   Peter Ho Chiung Ching.
Description: First edition. | Boca Raton, FL : CRC Press, 2022. | Series:
   Big data for industry 4.0 : challenges and applications | Includes
   bibliographical references and index.
Identifiers: LCCN 2021014164 (print) | LCCN 2021014165 (ebook) | ISBN
   9780367620066 (hbk) | ISBN 9780367620080 (pbk) | ISBN 9781003107477
   (ebk)
Subjects: LCSH: Wireless communication systems—Automatic control. |
   Wireless communication systems—Automatic control. | Machine learning.
Classification: LCC TK5105.2 .M33 2022 (print) | LCC TK5105.2 (ebook) |
   DDC 621.3840285/631—dc23
LC record available at https://lccn.loc.gov/2021014164
LC ebook record available at https://lccn.loc.gov/2021014165

ISBN: 978-0-367-62006-6 (hbk)
ISBN: 978-0-367-62008-0 (pbk)
ISBN: 978-1-003-10747-7 (ebk)

DOI: 10.1201/9781003107477

Typeset in Times
by Apex CoVantage, LLC

Contents

Preface

The design and development of automated approaches to improve the performance of wireless networks are considered among the challenging research issues in the field of wireless and mobile networking. The application of artificial intelligence (AI), machine learning (ML), and deep learning (DL) is relatively limited in the field of wireless networking systems and needs new models and methods to be developed to improve performance. Wireless network technologies such as the Internet of Things (IoT), Industry 4.0, Industrial Internet of Things (IIoT), VANET, and FANET-based applications demand data-driven approaches which involve complex mathematical models. These models can be automated and optimized using ML and DL techniques. AI-, ML-, and DL-based schemes are more adaptable to the wireless environment. These models provide an optimized way to reduce the complexity and overhead of the traditional tractable system models.

The large amount of data produced by wireless networks need to be stored and processed quickly to support real-time applications. This necessitates the attraction of data-driven approaches such as AI-, ML-, and DL-based schemes toward wireless communication and networking. Compared to traditional technologies, new technologies such as cyber-physical systems, cloud computing, virtualization, FANET, and VANET will have diverse service requirements and complicated system models that are harder to manage with conventional approaches properly. To cater to these needs, ML- and DL-based techniques can be employed in this domain to achieve automation. At present, automated learning algorithms in mobile wireless systems are in a growing phase, and the performance of these models needs to be optimized. This book aims to cover the state-of-the-art approaches in AI, ML, and DL for building intelligence in wireless and mobile networking systems.

It provides the latest advancements in the field of machine learning for wireless communications, encourages fruitful development on the challenges and prospects of this new research field. It provides a broad spectrum to understand the improvements in ML/DL that are motivated by the specific constraints posed by wireless communications.

Editors

K. Suganthi received her BE in computer science and engineering from Madras University, her master's in systems engineering and operations research, and her PhD in wireless sensor networks from the Anna University. She has been an assistant professor senior in the School of Electronics Engineering (SENSE) at Vellore Institute of Technology, Chennai campus, India, since 2016. She is the author of about 20 scientific publications in journals and international conferences. Her research interests include wireless sensor networks, the Internet of Things, data analytics, and artificial intelligence.

R. Karthik obtained his doctoral degree from Vellore Institute of Technology, India and his master's degree from Anna University, India. Currently, he serves as a senior assistant professor in the Research Center for Cyber Physical Systems, Vellore Institute of Technology, Chennai. His research interest includes deep learning, computer vision, digital image processing, and medical image analysis. He has published about 32 papers in peer-reviewed journals and conferences. He is an active reviewer for journals published by Elsevier, IEEE, Springer, and Nature.

G. Rajesh is an assistant professor in the Department of Information Technology of Anna University, Chennai, India. He completed his PhD from Anna University in wireless sensor networks and has approximately 12 years of teaching and research experience. His area of research interest includes wireless sensor networks and its Internet of Things applications, software engineering, and computational optimization. He has published more than 20 research papers in journals and conferences.

Peter Ho Chiung Ching received his PhD in information technology from the Faculty of Computing and Informatics, Multimedia University. His doctoral research work explored the performance evaluation of multimodal biometric systems using fusion techniques. Dr. Ho is a senior member of the Institute of Electrical and Electronics Engineers. Dr. Ho has published a number of peer-reviewed papers related to location intelligence, multimodal biometrics, action recognition, and text mining. He is currently an adjunct senior research fellow in the Department of Computing and Information Systems, School of Science and Technology, Sunway University.

1 Overview of Machine Learning and Deep Learning Approaches

Annie Johnson, Sundar Anand,
R. Karthik, and Ganesh Subramanian

CONTENTS

1.1 INTRODUCTION

In this age in which most platforms and services are automated, the network domain stands as no exception. Automation is applied to the various processes in the network life cycle that previously involved manual, time-consuming and unreliable

DOI: 10.1201/9781003107477-1

1

procedures. The network lifecycle consists of repeatable processes such as preparing, planning and designing, implementing, deploying, operating and optimizing the network. Traditional networks are slow and unresponsive as they are manually managed and hardware-centric. Therefore, to structure the networking systems better and intelligently control the cycle, software-defined networking (SDN) was introduced (1). The SDN approach enables a programmed and centrally controlled network that drastically improves the performance of the network.

It is beneficial to automate wireless networking systems as this would improve operational efficiency by reducing the number of network issues. As a result, the time involved in delivering solutions to those issues would also be minimal. Automation simplifies operations and makes the network cost-effective. These networks handle repetitive tasks with ease and are not susceptible to human errors. This establishes better control of the network and enables more innovations through the insights offered by network analytics. Automated networks are more resilient and experience lesser downtime. Hence, there has been a rise in the use of machine learning (ML) and deep learning (DL) techniques in network automation.

1.2 ML

ML is a subsection of artificial intelligence (AI) that equips computers to learn from data without having to explicitly program the learning algorithm. Developing an ML model capable of making accurate decisions consists of many stages beginning with the data collection phase. The data collected are usually split into two parts, namely, a training set that trains the ML model and a testing set used to determine the performance of the fully trained model. The data collected are then preprocessed in the data preparation stage. Then an appropriate algorithm to solve the problem at hand is determined. This is followed by the training phase during which the model identifies patterns and learns how to distinguish between the various input values provided. Once the model has been trained, it can be evaluated on a new set of data. These evaluation results are used to carry out parameter tuning and improve the performance of the model. Finally, the best network is used to make predictions. ML algorithms are useful as they can discover new patterns from massive amounts of data. These are several categories of ML algorithms that are classified based on multiple criteria.

Depending on whether or not these networks are trained with human supervision, ML algorithms are broadly classified into supervised learning, unsupervised learning and semi-supervised learning. The most widely used ML algorithms under each of these classes are discussed in the following sections.

1.2.1 Supervised Learning

ML models that utilize labeled data sets for training perform supervised learning. The algorithm relies on the output labels to form a relation between the input variable or the independent variable (X) and the output variable or the dependent variable (Y). The mapping function that denotes the relation between X and Y is represented as f. Supervised learning can be further classified into regression and classification problems based on the task performed by the algorithm.

Problems that involve the prediction of a numerical or continuous-valued output are known as regression problems. For example, if the price of a house is to be determined by leveraging features such as house plot area, number of floors, number of rooms and number of bathrooms, we would need input training data and the corresponding price labels. Using these data, a supervised learning model that predicts a numerical price value can be developed to solve this regression problem. Algorithms such as linear regression and logistic regression are popular regression-based ML algorithms that are used in supervised learning.

The second class of supervised learning problems is known as classification problems. Classification tasks involve mapping the test data to two or more categories. In these problems, the ML model is expected to provide only discrete output values that can be translated into one of the output classes. The most common type of problem that falls under this category is the image classification task. For instance, if images of cats and dogs had to be classified, then a supervised learning model for classification must be employed. Some well-known algorithms that fall under this category consist of k-nearest neighbors (KNNs), the Naïve Bayes model, Support Vector Machines (SVMs), decision trees and random forests.

1.2.1.1 Linear Regression

Linear regression is an algorithm that models the mapping function f as a linear function. It assumes that there is a linear relationship between the input and output data:

$$y = x\beta + \varepsilon .\tag{1.1}$$

In Equation 1.1, x is the independent variable that represents the input variable and y is the dependent variable that represents the output variable. The slope parameter, β, is termed as a regression coefficient, and ε is the error in predicting the y value. Here, y is depicted as a function (f) of x. The test data are entered into this linear function to predict the output value. Fig. 1.1 shows the prediction of y using a single input feature and the simple linear regression.

1.2.1.2 Logistic Regression

Input values are fed into the logistic regression model, which uses the logit function or the sigmoid function as shown in Equation 1.2 to produce output predictions that lie between 0 and 1. The logistic regression model can also be used to solve classification problems as the continuous-valued output values correspond to the probability of an instance being associated with a certain class:

$$P\left(y = \pm 1 | x, \beta\right) = \sigma\left(y\beta^{\mathrm{T}}x\right) = \frac{1}{1 + e^{-\left(y\beta^{\mathrm{T}}x\right)}}.\tag{1.2}$$

In Equation 1.2, $\beta = (\beta_0, \ldots, \beta_d)$ is a vector of dimension d, known as the model parameters, y is the class label which is ± 1 in the equation. The vector $x = (1, x_1, \ldots, x_d)$ are the covariates or input values (2).

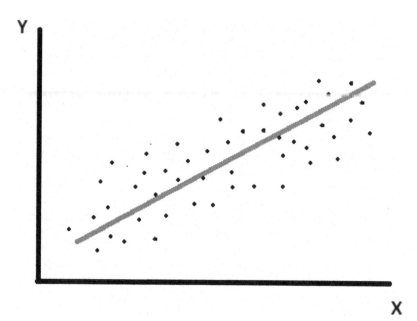

FIGURE 1.1   Simple Linear Regression.

1.2.1.3 KNNs

KNN is a classification model that labels the nearest patterns to a target pattern x. The class label is assigned to points based on a similarity measure in data space. Equation 1.3 defines the KNN for a binary classification problem:

$$f_{KNN}(x') = \begin{cases} 1 \, if \, \Sigma i \, N_k(x'), y \geq 0 \\ -1 \, if \, \Sigma i \, N_k(x'), y < 0 \end{cases}.$$

(1.3)

Here, y can either take the value of 1 or −1. $N_k(x')$ denotes the indices of K of the nearest patterns. K is the neighbourhood size. It is used to define the locality of the KNN. For smaller neighbourhood sizes ($K <= 2$), scattered patterns of different classes are obtained whereas, for larger neighbourhood sizes ($K > 19$), minority groups are ignored (3).

1.2.1.4 Naïve Bayes

The Naïve Bayes algorithm performs best on large data sets. This algorithm functions on the assumption that the various features of the data set are independent of each other. Then the Naïve Bayes model finds the probability of a new test sample belonging to a certain class and uses this parameter to perform classification. The model predicts the probability that a new sample, $x = (x_1, \ldots x_a)$, belongs to some class y, which is represented as $P(y \mid x)$. Here, x_i is the value of the attribute X_i, and $y \in \{1, \ldots, c\}$ is the value of the output class Y.

1.2.1.5 SVMs

SVMs use hyper-plane classifiers to separate the data points into their respective classes. The hyper-plane would be a point for or a one-dimensional data set, a line for a two-dimensional data set, a plane for a three-dimensional data set and a hyper-plane for any data set having a dimension higher than three. A linearly separable SVM classifier is denoted by Equation 1.4 (4):

$$ax + by + c = 0 . \tag{1.4}$$

Here, (x,y) are the data points. The slope of the linear classifier is given by (a/b), and the intercept term is (c/b).

1.2.1.6 Decision Trees and Random Forests

One of the most basic classifiers, a decision tree performs two tasks, learning and classification. Based on the training data set, the decision tree learns the split criterion. This phase is known as the learning phase. The phase that follows the training phase is the classification phase, during which the test data are classified using the trained tree. The tree has a structure resembling a flow chart and consists of three parts known as the leaf node, the branch and the internal nodes. Each branch of the tree represents the output obtained on a test condition. The bottommost node that holds the final predicted output class is called the leaf node. A decision tree is one of the simplest machine learning classifiers and is easy to comprehend. However, one of the challenges faced by the decision tree algorithm is that it is more likely to overfit the data. Therefore, it is a weak classifier. Hence, many decision trees are combined to form a stronger classifier known as a random forest. The random forest is an ensemble model that provides its final classification by choosing the most popular class predicted by the decision trees for a data sample as the final classification for that particular data sample (5).

1.2.2 Unsupervised Learning

Unsupervised learning involves the discovery of previously unknown patterns from unlabeled data. Unlike supervised algorithms, unsupervised algorithms can help address a wider range of problems as it is easier to obtain unlabeled data. Unsupervised ML algorithms can fall under three types, which include clustering algorithms, visualization and dimensionality reduction algorithms and association rule learning algorithms. This classification is based on the type of task performed by the algorithm.

Clustering algorithms find a structure in the uncategorized data. Similar data points are grouped together. For instance, segregating consumers into groups, with the help of clustering models, would help businesses target their customers better and get the best return on investment. k-means and hierarchical cluster analysis (HCA) are common clustering algorithms.

Visualization and dimensionality reduction algorithms perform related tasks. A visualization algorithm is used to model and plot unstructured data as a two- or three-dimensional representation. This helps in identifying unsuspected patterns in the data. For example, visualizing the most spoken languages in the world would require such visualization algorithms. Dimensionality reduction is a technique that merges correlated features into a single feature and, as a result, simplifies the available data without losing too much information. The mileage of a car may be correlated with the age of the car. Therefore, using dimensionality reduction or feature extraction these correlated features can be merged into a single feature named the wear and tear of the car. By utilizing this technique, the model will train faster and lesser memory space is required to hold the data. Principal component analysis (PCA) and kernel PCA are popular visualization and dimensionality reduction algorithms.

The goal of association rule learning algorithms is to explore large data files and discover interesting patterns and new relations between the various features of the data. Association rule learning algorithms may be applied in supermarkets, whereby the algorithm may reveal that people who buy bread are more likely to buy bread spreads also in that purchase. Therefore, it would be ideal to place these two products next to each other. Some association rule learning algorithms include apriori and ECLAT.

1.2.2.1 k-Means

k-means is one of the primitive clustering algorithms that can be used for grouping problems. A group of randomly selected centroids are used as the initial centres of k clusters. k represents the number of clusters required. This parameter is also set before running the k-means algorithm. The algorithm then performs a series of calculations that influence the new set of k centroids for the next iteration. After completing the defined number of iterations, k clusters are obtained. k-means is computationally faster than HCA and produces tighter and more spherical clusters. However, it is challenging to determine the perfect k value (6).

1.2.2.2 HCA

HCA is a clustering algorithm that can be classified as agglomerative HCA and divisive HCA. In agglomerative clustering, each input data sample is assumed to be an individual cluster. Similar clusters then merge into one another until 'k' distinct clusters are obtained. This happens after every iteration is complete. The clusters are grouped based on a proximity matrix that is updated every time the iteration is complete. The divisive HCA algorithm initially considers all the data points to belong to a single cluster. Data points that are not similar are then separated from the cluster. This algorithm is not as widely used as the agglomerative clustering technique.

1.2.2.3 PCA

The PCA dimensionality reduction technique is used to convert data sets having a large number of features into ones with fewer features. It can only be applied to linear data sets which are data sets that are linearly separable. The data set, having fewer attributes, would still contain most of the information. Data sets having a smaller number of features are easier to explore and analyse. ML algorithms train faster on the data sets that have undergone dimensionality reduction. In PCA, to ensure that every variable has an equal weight in contributing to the analysis, the variables are initially standardized. Then, the covariance matrix of the data set is constructed. The principal components of the data are determined from the eigenvalues and eigenvectors of the covariance matrix. The principal components are the updated set of features that can be represented as a linear combination of the original set of features. This technique greatly increases the classification accuracy of a model (7).

1.2.2.4 Kernel PCA

Kernel PCA is an extension of PCA. It is a dimensionality reduction technique that can be applied to nonlinear data sets. A kernel function is used to project the data set into a feature space where it is linearly separable. The kernel function acts as a replacement to the covariance matrix calculated in PCA. It is used to calculate the eigenvalues and the eigenvectors that are required to obtain the principal components of a given data set. The most commonly used kernels are the polynomial kernel and the Gaussian kernel. Polynomial kernels are used for data sets modelled with non-linear decision boundaries that are polynomial in shape, whereas, for data points that are distinguished based on the distance from a centre point, Gaussian kernels would be the preferred kernel function. Kernel PCA has an advantage over PCA as real-time data are more likely to be non-linear in nature (7).

1.2.2.5 Apriori

The apriori algorithm is a popular algorithm used in data mining to determine the relationship between different products. These relations are termed as association rules. The various items in the data set are mined and the set of items or the item set that occurs most frequently is determined using the apriori algorithm. The main factors that are used in the apriori algorithm are support, confidence and lift. The support is the probability that two items in the data set (A and B) occur together. Confidence is the conditional probability of B, given A. Lift is the ratio of support to confidence. Using these parameters and a breadth-first search approach, the apriori algorithm can determine the frequent item sets in the data set (8).

1.2.2.6 Equivalence Class Transformation Algorithm (ECLAT)

On the other hand, the ECLAT utilizes a depth-first search approach to determine the frequent item sets in a given data set. The input to this algorithm is a transaction database. A set of transactions is collectively defined as a transaction database and a transaction is an itemset. The algorithm discovers frequent item sets and association rules from the transaction database. As the ECLAT algorithm uses a depth-first search in the database, it is faster than the apriori algorithm and has a lower memory requirement (8).

1.2.3 SEMI-SUPERVISED LEARNING

In semi-supervised learning, algorithms can handle a combination predominantly consisting of unlabelled data and a much smaller amount of labelled data. This is particularly useful in the medical field as it usually takes a lot of time and the expertise of medical professionals to label medical scan images. Semi-supervised learning algorithms would require only a few labelled images, thus saving a lot of time and effort.

1.2.4 Analysis of ML

ML algorithms are widely used in the wireless networking domain. For instance, logistic regression models are used in determining the probability of failure of a network or a process. This is a regression problem. Classification problems such as predicting root-to-local (R2L) or denial-of-service (DoS) attacks in the networking domain can also leverage ML algorithms (9). The ML-based solutions to networking problems can also make use of feature engineering techniques like dimensionality reduction. Hence, ML in the networking domain can be used to speed up and efficiently perform fundamental networking tasks including traffic prediction, network security and packet routing. In spite of the multiple advantages of ML in networking, ML algorithms still have limitations and face challenges. ML algorithms require hand-picked features to train the network, and this tends to influence the performance of the model. Another major drawback of ML algorithms is that these algorithms require a huge amount of data for training. Fewer available data give rise to the problem of overfitting. More training data could also mean higher computation costs. Hence, DL models were introduced to overcome these challenges.

1.3 DL

DL is another branch of AI. Unlike ML, DL doesn't treat all the features equally. DL first learns which all features significantly impact the outcome and based on that the DL creates a combination of all features for the learning process. This property of DL demands a lot of data. A DL model has at least one or more hidden layers. The hidden layers fall between the input and output layers. Hidden layers are intermediate layers through which the DL algorithm learns which combination of features can be used to get the best consistent results. DL is widely used in various supervised classification and regression problems. The training of the deep learning algorithms happens via back propagation, whereby the algorithm learns the parameters for each layer from the immediate next layer and so on. Some of the well-known DL algorithms are recurrent neural networks (RNNs), convolution neural networks (CNNs) and general adversarial neural networks (GANs). Generally, these models have many different data-processing blocks before the hidden layers. Some of the commonly used blocks are convolution, pooling and normalization.

The convolution block use kernels (or filters) to convolute multiple features at a time depending on the kernel size to get the spatial information about the data.

The pooling block is used to decrease the feature set size by either taking average or max of multiple features. This helps increase the computation speed of the algorithm and, at the same time, preserve the information.

Normalization is used to normalize the data in a feature. This is because due to multiple processing steps, the data may change significantly, and if one feature has relatively higher numbers than another feature, then the feature with a higher number dominant the results. To avoid this, we normalize the data across features so that all features are weighted equally before they enter into the hidden layers.

1.3.1 CNNs

CNNs use convolution block as one of the major functions to get the most prominent combination of features to get the results. This approach enables the algorithm to successfully capture the temporal and special dependencies between different features. The architecture of CNN facilitates the reduction of the size of the features which are easier to process, and it gets the results without losing any information.

1.3.2 RNNs

RNNs learn just like CNNs, but RNNs also remember the learning from prior inputs (10). This context-based learning approach from RNNs makes them suitable for any sequential data as the model can remember the previous inputs and the parameters learnt from them. Hence, this architecture is one of the best choices to make when dealing with series data as this model uses the data from the past to predict the present output values.

1.3.3 GANs

GANs are used for data augmentation. GANs can produce new data points with the probability distribution of the existing data points over N dimensional space. The GAN model has two parts: (1) generator and (2) discriminator. The generator is used to create fake data points in addition to the existing data points based on random inputs, and the discriminator is used to classify the fake points from the existing data points. This process is repeated by updating the weights of the generator such that it increases the classification error and the weights of the discriminator such that it decreases the classification error until we get the fake points to have almost the same distribution of the original existing data points. In this way, the GAN model is able to generate new data points which have almost same probability distribution as the existing data points.

1.3.4 Analysis of DL

DL is preferred over ML because DL automates the feature selection, and the extraction process is automated as well. In ML, the features are hand-picked manually and fed to the model. DL removes this process with the help of blocks and hidden layers, whereby the model learns what combination of the feature works well for the data set considered. But at the same time, DL also has its downside. To run a DL model, a huge amount of data is required. The amount of data is proportional to the feature extraction efficiency of the DL model. So if the data set size is small, then ML algorithms perform better than DL algorithms.

1.4 CONCLUSION

ML and DL models have greatly influenced the way automation happens in today's world. The development of these algorithms has enabled automation in every field, including networks for wireless devices. By implementing ML or DL in a wireless networking domain, various processes which involved manual, time-consuming and unreliable processes are now being more refined and automated. This way a lot of manual errors and time delays are rectified. By removing manual works and automating them, the operational efficiency to find and resolve network issues is minimal and cost-effective. Hence, the evolution of these ML and DL architectures have greatly contributed to the development of the modern network domain.

REFERENCES

1. Xie, Junfeng et al. (2019). A Survey of Machine Learning Techniques Applied to Software Defined Networking (SDN): Research Issues and Challenges. *IEEE Communications Surveys & Tutorials*, 21(1), 393–430. Institute of Electrical and Electronics Engineers (IEEE), doi:10.1109/comst.2018.2866942.
2. Bonte, Charlotte, and Frederik Vercauteren. (2018). Privacy-Preserving Logistic Regression Training. *BMC Medical Genomics*, 11(S4). Springer Science and Business Media LLC, doi:10.1186/s12920-018-0398-y.
3. Kramer, Oliver. (2013). K-Nearest Neighbors. *Dimensionality Reduction with Unsupervised Nearest Neighbors*, 13–23, doi:10.1007/978-3-642-38652-7_2. Accessed 27 December 2020.
4. Taher, Kazi Abu et al. (2019). Network Intrusion Detection Using Supervised Machine Learning Technique with Feature Selection. *2019 International Conference on Robotics, Electrical and Signal Processing Techniques (ICREST)*, doi:10.1109/icrest.2019.8644161.
5. Lan, Ting et al. (2020). A Comparative Study of Decision Tree, Random Forest, and Convolutional Neural Network for Spread-F Identification. *Advances in Space Research*, 65(8), 2052–2061. Elsevier BV, doi:10.1016/j.asr.2020.01.036.
6. Fränti, Pasi, and Sami Sieranoja. (2018). K-Means Properties on Six Clustering Benchmark Datasets. *Applied Intelligence*, 48(12), 4743–4759. Springer Science and Business Media LLC, doi:10.1007/s10489-018-1238-7.
7. Datta, Aloke et al. (2017). PCA, Kernel PCA and Dimensionality Reduction in Hyperspectral Images. *Advances in Principal Component Analysis*, 19–46, doi:10.1007/978-981-10-6704-4_2.
8. Robu, Vlad, and Vitor Duarte Dos Santos. (2019). Mining Frequent Patterns in Data Using Apriori and Eclat: A Comparison of the Algorithm Performance and Association Rule Generation. *2019 6th International Conference on Systems and Informatics (ICSAI)*, doi:10.1109/icsai48974.2019.9010367.
9. Boutaba, Raouf et al. (2018). A Comprehensive Survey on Machine Learning for Networking: Evolution, Applications and Research Opportunities. *Journal of Internet Services and Applications*, 9(1). Springer Science and Business Media LLC, doi:10.1186/s13174-018-0087-2.
10. Mikolov, Tomas et al. (2011). Extensions of Recurrent Neural Network Language Model. *2011 IEEE International Conference on Acoustics, Speech and Signal Processing (ICASSP)*, doi:10.1109/icassp.2011.5947611.

2 ML and DL Approaches for Intelligent Wireless Sensor Networks

Bhanu Chander

CONTENTS

DOI: 10.1201/9781003107477-2

2.1 INTRODUCTION

Due to the tremendous developments in various domains, the upcoming smart communities insist on a vast amount of information for monitoring or supervising different human-related behaviors. Until recently, wireless sensor networks (WSNs) have had enough capability to provide this statistical information with low-cost and less intelligent systems. Although the sensor nodes contain minimal resources and the requested deep-seated performances of WSNs, they are better than before; moreover, there is a great demand for smart communities that overcome the disputes created by the global world (1–3). We can easily find numerous kinds of literature with various techniques for efficient WSN deployments to take full advantage of network duration and performances. However, most of them took on several issues individually. The authors afford efficient management approaches dependent on existing disputes; however, the authors failed to suggest innovative tools or methodologies. To tackle this difficulty in a global behavior, intelligent WSNs are proposed is introduced by Yang et al. and Alsheikh et al. (1, 2).

WSNs are a networking and computing field containing small, intelligent, and low-cost sensor nodes. Moreover, these nodes help monitor, analyze, and have power over physical environmental variables such as humidity, pressure, pollution status, vibration, temperature, movement, signals, and light, among others. Whatever wireless-based technologies we are enjoying today have not come in a straight way; there was tremendous research work hidden in WSNs technology innovation. The initial steps of WSNs are placed in 490 BC when the Greeks defeated the Persians (the Battle of Marathon), and the message was carried nearly 44 km to its destination, Marathon, which was run by a soldier. Today, the same message is delivered in a flash with wireless communications. After the Marathon message delivery, drums, smoke signals, semaphore, and heliographs were used for message transmission. The first breakthrough in long-established distance communication took place in 1800 with the implementation of electrical circuits to transform signals over wires. After this, in the early 1830s, famous researchers Cooke and Wheatstone verified a telegraph method through five magnetic needles, whereby electric power forced the needles to point at the letters as well as the numbers that shaped the message's meaning.

Moreover, the main object of electronic wireless communication arose when Maxwell declared all electromagnetic waves move at the speed of light; moreover, he also verified the association struck between electrical energy and magnetism (4–8). Marconi, in 1985, made a handy arrangement for signal diffusion over one-half mile. After this, a breakthrough was made by famous scientist Edwin Armstrong that included the fundamentals for today's necessary discoveries: frequency modulation, regeneration, and super-heterodyning. With these, Armstrong created a healthy stage for advancing wireless communications, television, mobile communication, and spread spectrum. At the end of the 19th century, the word *wireless* turned into *wireless telegraphy*, which is well known as radio. Most wireless communication utilizes radio frequencies; however, optical, infrared, acoustic, and magnetic frequencies can be used.

Systems also some well-defined permitted wireless communication. A wireless scheme incorporates a broad series of fixed, mobile, and handy applications. Designing a wireless scheme entails equivalent disputes as a wired scheme and antennas plus propagation channels. In general, a message holds facts that a sender wishes only the receiver to be familiar with. In most cases, messages in the form of yes or no, while it can be complex material like video or voice. In WSNs, the communication process starts with information exchange or transfer, and it may be any data. Each sensor node contains an analog-to-digital converter to transform the analog signals into bits. For instance, the 8-bit ASCII code for the symbol "2" is 00110010. Complementary bits added to the code notice and correct errors, which is acknowledged as channel coding. This channel coding allocates the recipient to the correct errors provoked via the channel. The modulator maps the channel encoder end-result to an analog signal apt for broadcast into the channel. An antenna extends the indications into the channel at one end, and a new antenna collects the other end's indications. A channel is a pathway taken by the broadcasted signal to the receiver. Signals turn indistinct, strident, and attenuated in the channel. The demodulator translates the acknowledged analog signal into a digital signal that nourishes the channel decoder and others ahead of receipt by the recipient (3–8). Active signal exposures occur when the signal power surpasses the recipient threshold; noise, along with interference, do not provoke errors that cannot be approved.

The need for communication services of WSNs in various situations has become a tricky concern, particularly in the latest technologies like the Internet of Things (IoT), wireless health care management, and mobile and smart communications. In such smart setups, WSNs are likely to offer a ubiquitous arrangement of allied devices that make it possible support to information and sensing, as well as communication, structures (2). The primary purpose of WSNs are fashioned by a collection of minute plus superficial communication nodes with power sovereignty, condensed processing ability, and, notably, accurate sensing talent. In general, the temperament, as well as aspects, of the WSN circumstances hold with a significant quantity of sensor nodes coupled to a central node, which is also known as a "high energy communication node," tied to a vast quantity of precise applications and offer numerous disputes in WSN design region in execution.

2.2 INTELLIGENT WSNS

As a result of continuous progress and various application services for machines, devices, and things in WSNs, now it is feasible for wireless communication and network skills to move one more step ahead as a talented idea "fifth generation" (5G) wireless network. With this innovation, WSNs now have more command of interlocking, a great figure of smart devices with intelligence and reconfiguring capacity. The more important point is that they elegantly formulate conclusions with a brain like a system for advanced intelligence. That is why it is like a tonic for many complicated applications in a variety of domains like remote sensing, space technology, health management, smart cities and grids, intelligent transportation, defense field tracking and monitoring, vehicle repair and performance checking, and many more, which significantly improve our quality of life (2–5).

However, the preceding mentions one side of the coin; on the other side of the coin is applying intelligent wireless networks to raise numerous key disputes. For example, an increased number of smart devices produce a huge amount of raw data records. However, problems arise from data gathering, processing, analyzing, accessing, and accurately using them. Besides, the tremendous service necessities smart wireless procedures; plus, complex/dynamic situations, its devices, its services, and its applications are still not well groomed enough to deal with optimized physical layer aims, complex decision-making, and well-organized resource managing assignments in upcoming wireless networks (2–6). Machine learning (ML) models have to get attention from researchers as a promising solution to gain an advantage from intelligent nodes and minimize challenges.

During the past decades, ML has been recognized as one of the promising, authoritative artificial intelligence (AI) frameworks, and it has expansively incorporated into image processing, social business, signal and natural language processing, health care, and behavior analysis. It mostly uses statistical models for analysis, preprocessing, and observations by detecting and describing the underlying structures (6–9).

Incorporating intelligence into WSNs must cover the subsequent perception ability, reconfigurability, and learning capacity. In perception ability, WSN environment awareness is an essential characteristic in intelligent wireless-based connections (3). It makes wireless functions of a device settle into its location and potentially best use the offered spectrum resources. Perception ability presented by spectrum sensing, and some highly developed spectrum-sensing methods manage a mixture of situations. As spectrum sensing demands resources at the sensing nodes, the resourceful arrangement of spectrum sensing has been employed for the stability of power, time, and bandwidth, as well as sensing and transmission. To adapt to dynamic changes of the surrounding conditions, intelligent WSNs require reconfigurable ability, and it's achieved by dynamic spectrum access and the optimization of operational parameters. The physical layer's key reconfigurable constraints contain waveform, power allocation, time slot, modulation, and frequency band (5–9).

Finally, the learning potentiality allows wireless communication devices to autonomously become skilled at optimal models according to the corresponding wireless environment. To accomplish high performance, like throughput, proper power usages, and comfort, certain limits, such as quality of service (QoS), need diverse optimization models consisting of graph-based and market-based models developed. The main disputes on the issue, counting damaged information, and real-time supplies, along with complexity restrictions, must consider. With a large number of wireless devices requiring a large quantity of energy, energy effectiveness also becomes vital for dynamic spectrum access as well as resource optimization (5–9). As a result, intelligent WSNs have garnered greater attention than before.

This chapter pays attention to exploring employing ML approaches to solve the critical disputes in intelligent WSNs. Intelligent wireless communications, whereby each of the following sections establishes the necessary algorithms of ML and then discusses quite a few cases of applying such techniques or algorithms for intelligent WSNs.

2.3 NEED FOR DATA PREPROCESSING

WSNs collect a considerable amount of raw data from assigned regions continually. Transforming this raw data to the central node (base station) consumes a high amount of a sensor node that holds restricted resources. Moreover, the collected raw data contaminated with noise, error values, incorrect nodes information, and the like; if it is transformed into an expert system for analysis purposes without any modifications, the expert system may give false outcomes. Moreover, the raw or unprocessed data may not suitable for straight use by the learning models or expert systems. Moreover, in most cases, data are spread over multiple database sources, and data include irrelevant or insignificant factors that will not help to solve the issue. More important, raw data do not imitate the target problem's real activities and hold parameters with different scales (3–5). Hence, there is a huge demand for transforming or preprocessing the collected data or raw data into an appropriate arrangement before feeding them into an expert system or ML model. Preprocessing highlights that the collected data have to process before training. Data reduction, data transformation, data cleansing, and data integration are well-known approaches in preprocessing.

2.3.1 DATA REDUCTION

Data reduction is the elimination or abolishment of irrelevant information concerning the objective of the assignment. We must choose relevant instances and system properties for the best training. Dissimilar feature selection models or the features that represent the data, such as extraction models applied for data reduction.

2.3.2 DATA TRANSFORMATION

Data transformation involves discovering an original form of data that is more appropriate to instruct the learning models closely related to normalization. Most of the data transformation techniques use statistical models by scaling them to fall within a slighter preference. Making the designed constraints to a similar level by standardizing the physical layer constraints to (0, 1) will computationally accelerate the learning and forecast progression of the models.

2.3.3 DATA CLEANING

Data cleaning recognizes the anomaly observation; some corrected, and some removed from the data set. Deleting these values from the collected data can lead to a newfangled, more seamless prediction form—numerous outlier- or anomaly-exposure techniques for better decision-making.

2.3.4 DATA INTEGRATION

Data integration is applied when the data travel through different databases or files. This problem mostly occurs in complex web-based production systems and IoT and cyber-physical systems.

2.4 ML FOR INTELLIGENT WIRELESS COMMUNICATIONS

WSNs consist of numerous small, autonomous, low-power, limited-bandwidth, and low-cost sensor nodes. Mostly WSNs have been employed in dynamic environments that change quickly. Thus, a WSN is one of the most capable pieces of equipment for several real-time applications because of its size, productivity, and effortless deployable feature. Here, sensor nodes' goal is to supervise a field of interest, gather data, and collaborate to forward raw sensed data to the back-end unit's entitled base station or sink node for further dispensation and analysis. By analyzing with expert systems, upcoming events can be explored. In a WSN, sensor nodes outfitted with a mixture of sensors like pressure, thermal, weather, acoustic, optical, and chemical sensors. Depending on the application, a single WSN contains hundreds to thousands of nodes, and some of them explicitly build for particular event measures. So supervision of such large networks with different nodes requires a well-organized and scalable algorithm.

Nevertheless, implementing suitable and practical approaches to dissimilar applications is not an easy task. Implementing perfect, effective models for dissimilar WSN applications is not an easy task (3–8). Designers must address fundamental concerns like security, localization, data aggregation, node clustering, fault detection, reliability, and preferred event scheduling.

Moreover, unexpected internal (by design) or external (by environmental conditions) positions of WSNs may adjust dynamically. Moreover, it significantly affects the network coverage, delay transmission, routing tactics, QoS, link quality, and the like. For these reasons, it may call for depreciating, not essential refurbish of the setup of the extremely dynamic nature. However, the long-established techniques for the WSNs are programmed as well as planned. As a result, the setup does not work as it should for the dynamic atmosphere. To settle such circumstances, WSNs often assume ML proficiency for abolishing the need for a redundant revamp. ML also motivates many handy solutions that exploit resource exploitation and expand lifetime (3–9).

ML is the subpart of AI, and it received attention between 1950 and 1952. ML strengthened most of the computational algorithms. ML defines a process with the impressive capability of involuntarily learning and, developing from the knowledge or skill, takes steps without being explicitly programmed or planned. Since ML's introduction, it has employed in several field assignments like data classification, clustering, and regression in a wide variety of applications such as anomaly detection, computer vision, health care management, spatial data analysis, machinery performance tracking, intricate networks design, speech recognition, fraud detection, and bioinformatics, among others (3–9). ML's algorithms and practices come from various miscellaneous fields such as statistics, neuroscience, computer science, and mathematics. ML makes our computing methods more and more resourceful, consistent, and commercial. ML formulates the best procedures by analyzing even more multifarious data repeatedly, quickly, and more precisely. ML models' success depends on their capability to offer comprehensive solutions through a structural design that improves model performances by learning consistent data pattern structures. ML is defined as "the development of computer models for learning processes that provide solutions to the problem of knowledge acquisition and enhance the performance of developed systems" (2) or "the adoption of computational methods for improving machine performance by detecting and describing consistencies and patterns in training data" (6). Incorporating these definitions into WSNs will advance the performance of sensor nodes on provided assignments without any reprogramming. More specifically, ML is necessary for WSN applications for the following motivations:

1. Most of the time, sensor nodes are employed in hazardous and dynamic locations; the internal and external factors location also changes dynamically. So it needs to develop perfect localization precisely and quickly with the assistance of ML models.
2. Sensor nodes continuously collect raw sensed data from the surrounding environment, but broadcasting complete data to the centralized node will increase transmission complexity. ML models extract the features (dimensionality reduction) that enhance decision-making at cluster head (CH) and non-CH member levels.
3. In some special applications, sensor nodes are also employed for acquiring or learning actual knowledge from dangerous environments (e.g., underwater level, in-depth forest weather forecasting, etc.). However, due to the unanticipated, sudden changes in those situations, designed situations may not work as expected. Here, system designers utilize the capabilities of

vigorous ML models to make nodes calibrate themselves to newly acquired knowledge.

4. Target area coverage is another big issue while deploying WSNs; choosing the best possible quantity of sensor nodes to wrap the region is effortlessly achieved by ML practices.

5. ML models are utilized to separate the defective or damaged sensor nodes from standard nodes and progress the efficiency of the network.

6. WSNs are deployed in challenging environments where humans find it difficult to develop a regular network with perfect mathematical models to explicate a system's activities. However, some tasks in WSNs solve with standard mathematical representations, but still, there is a huge demand for complex models to solve them. Under such conditions, ML provides low-complexity estimations for the system model.

7. Energy-harvesting offers a self-powered plus long-term continuation for the WSNs position in the unsympathetic location. ML models progress the performance of WSNs to expect the quantity of power to be produced in a specific time slot.

8. The transmission of sensed data with an accurate routing technique boosts the network's existence. Similarly, the sensor network's dynamic activities need energetic routing methods to improve structural performance, which will be done by the appropriate ML models.

9. At present, WSNs are incorporated with other technologies like the IoT, health care management, cyber-physical systems, the agriculture sector, and machine-to-machine communication. Here, ML models are vital to excavate the different concepts required to complete the AI tasks through partial human involvement.

This chapter's sections mostly focused on ML for intelligent wireless communications. The source of WSNs concerning surrounding environments need to be identified because of potentiality, reconfigurability-characterized power, frequency length, processing capability, memory storages, and the like. Associations among these aspects in the light of how they contact the overall method efficacy are still not clearly understood. Modern-day ML methods will find good opportunities in this particular application. ML provides or designs processes that lead the network's remodulation to view the surrounding environmental conditions and confirm the outcome by taking full advantage of all the accessible resources. In other words, the learning potentialities allow wireless devices to learn without optimally settling into the wireless environment.

2.5 ML APPROACHES

ML models try to learn or gain knowledge of task (T) based on a particular experience (incident; E), the objective of which is to advances the performance of the job calculated with an accurate performance metric (P) through exploiting experience E. We employ or state T, E, and P ML models categorized into supervised, unsupervised, semi-supervised, and reinforcement. We discuss each of these categories with up-to-date models and their applications. Moreover, we try to explore the latest techniques like neural network-based deep learning (DL) for enhancing intelligent communications. First of all, different parts of ML algorithms found in their functionalities, like support vector machines (SVM), clustering, statistically improved models, Bayesian learning, principal component analysis, and Q-learning, are described (Figures 2.1 and 2.2).

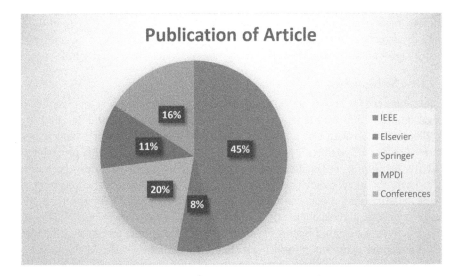

FIGURE 2.1   Pie Chart of the Percentage of Articles Published.

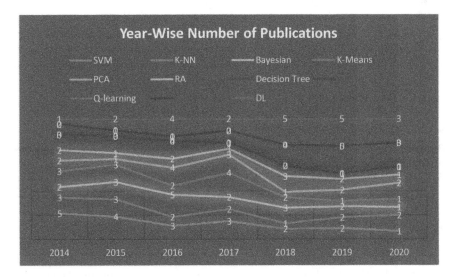

FIGURE 2.2   Year-Wise Publication of Various Algorithms for Intelligent WSNs.

2.5.1 SUPERVISED LEARNING

In supervised learning, tasks able to gain knowledge from the data sample provided by some external administrator. Here, the administrator prepares a labeled trained data set that holds both predefined inputs and known outputs. A supervised model is employed to symbolize the learned correlation among the input, output, and system factors. Every training instance holds a pair of input and unexpected output; the purpose is to learn a task that forecasts acceptable output for any input. We try to explore the foremost supervised learning models that broadly utilize to crack numerous disputes in WSNs. In truth, most of the complex issues like event detection, media access control, data integrity, object targeting, query processing, QoS, intrusion detection, localization, and security protocols are extensively solved by supervised learning approaches.

2.5.1.1 Decision Tree

A decision tree (DT) is one of the prominent supervised ML models for classification based on predicting data labels by iterating the input data through a learning tree. Furthermore, a set of if–else set of laws is made the model's choice or sorting to boost the readability. DT utilizes a predicted class or marks throughout this procedure by building a training replica founded on decision rules incidental from the training data. Here, the most critical benefit of DTs is that they are incredibly apparent and lessen indistinctness in decision-making and consent for a wide-ranging analysis. DT has heavy usage and excellent adoptive capabilities in solving various WSN issues, like discovering essential features such as anomaly detection, loss rate, path collection, mean time to failure, refurbish, and data aggregation.

2.5.1.2 Random Forest

Random forest (RF) is well known as an advanced version of DT, with a collection of trees, and each of them offers a classification. RF contains two phases, building an RF classifier and forecasting consequences, that powerfully work on varied, large quantity data sets. RF correctly identifies the missing values and existing models facing key challenges owed to a curse of dimensionality and interconnected data. RF algorithm has been widely employed for various applications and to resolve various concerns in WSNs like coverage, localization, data aggregation, and Medium Access Control (MAC) protocol.

2.5.1.3 Artificial Neural Networks

Artificial neural networks (ANNs) work like a human neuron system, with many neuron units for the perfect categorization of raw data. ANNs catalog composite, non-linear data sets without any difficulties; plus, there are no limitations for the inputs similar to the remaining classification process. ANNs can help develop the effectiveness of a variety of issues in WSNs, including intrusion detection, localization, and data preprocessing, to discover faulty sensor nodes, routing, and congestion control.

2.5.1.4 SVMs

In SVMs, the classifier detects a most favorable hyperplane to sort the data set with labeled training examples. Furthermore, most of the data points or examples are not required if the borderline predicted or decided—the data points used to locate the authorized borderline support vectors. SVM gives the top categorization from a prearranged set of data. SVM models are extensively used in WSNs for congestion control, anomaly detection, localization, routing, and connectivity issues.

2.5.1.5 Bayesian Models

A Bayesian ML model, compared to other ML models, provides accurate classification based on statistical learning techniques. The Bayesian interface requires little quantity of training samples. Bayesian MLs train or find out relationships among data points through conditional independence using statistical methods. It employs different possibility assignments for different variables of class nodes. In recent times, numerous WSNs' troubles are answered based on the Bayesian learning policies toward improving system efficiency. Most Bayesian models adjust for the possible distribution leading to efficiently finding out uncertain perceptions without overfitting. Here, the main thing is to utilize existing knowledge and then modernize the previous beliefs into subsequent beliefs. Forecasting environment conditions with incomplete or inappropriate data sets. Bayesian MLs broadly are applied in WSN applications, such as coverage issues, event detection, localization, target tracking, path selection, routing, anomaly, and event detection.

2.5.1.6 k-Nearest Neighbor

k-nearest neighbor (KNN) categorizes data depending on the distance of the samples exactly in training plus test samples. It uses numerous distance measures like Manhattan distance, Minkowski distance, Hamming distance, Euclidian distance, Chebyshev distance, Mahalanobis distance, and Canberra distance. KNN does not require high computational demand, as the purpose is to calculate comparative confined points. This factor fixed by the interconnected analysis of adjacent nodes formulates KNN as an appropriate scattered learning model for WSNs. The difficulty of the KNN models depends on the size of the input data set and the most favorable performance on similar-level data. KNN accurately finds the potential misplaced values from the feature space as well as shrink the dimensionality. KNN is mostly applied to data aggregation, intrusion detection, faulty nodes, and query-processing subsystems.

2.5.2 Unsupervised Learning

Unsupervised models deal with input data, and there is no output that bonds with the inputs. Models strive to dig out the real correlation from the existing records. The key uses of unsupervised learning (clustering models, dimensionality reduction models) in WSNs are anomaly detection, intrusion detection, QoS, routing, and data aggregation.

2.5.2.1 k-means Clustering

k-means clustering considers a k number of random positions and builds an absolute number of clusters from the given data. After forming clusters by considering every data point, a novel centroid for every cluster was redesigned. After each epoch or iteration, the centroid changes location and replicates the model until there are no more alterations in the centroid of each cluster. It is the simplest clustering; moreover, it is valuable in WSNs for discovering the most advantageous CHs for routing the information toward to sink node. Decision-making, data collection, and feature extraction are some of the applications for k-means clustering in WSNs.

2.5.2.2 Hierarchical Clustering

In hierarchical clustering approaches, there is no need for prior information, for cluster numbers divide the data into distinct parts and clusters similar objects as groups with prearranged top-down or else bottom-up array. In top-down or divisive clustering, a single outsized panel splits recursively in anticipation of one cluster for every observation. Bottom-up or agglomerative clustering is based on density functions; every observation consigns to the nearest CH. It is used to explain a range of problems in WSNs, data aggregation, management, mobile sink, and energy harvesting.

2.5.2.3 Fuzzy-c-Means Clustering

Fuzzy c-means (FCM) clustering was defined and developed by the famous scientist Bezdek in 1981 by using fuzzy set theory. In this model, clusters are described or designed based on similarities such as strength, distance, and connectivity. Furthermore, these similarities are picked based on applications and data sets. It provides the best results compared to k-means clustering, but the complications of FCM are greater than all other existing cluster models. FCM practices are used to resolve numerous issues in WSNs like localization, connectivity, and mobile sinks.

2.5.2.4 Singular Value Decomposition

Singular value decomposition (SVD) is mainly applied to dimensionality reduction in matrix factorization, representing a matrix in the matrix's product. SVD is capable of powerfully decreasing the data dimensions of the particular feature space. SVD assurances an adequate low-rank demonstration of the data. SVD uses WSNs to tackle diverse topics similar to routing, classification, path selection, frequency modulation, and data aggregation.

2.5.2.5 Principal Component Analysis

Principal component analysis (PCA) is a multivariate scheme employed for dimensionality diminution and data compression. Specifically, it extorts essential information from data and presents it as a group of new variables entitled as principal components. In detail, PCS aggregates every data point into principal components. Higher components hold the uppermost discrepancy manner of the data and release the least priority information from the feature space, decreasing dimensionality. For instance, in WSNs, nodes collect massive raw data; transmitting this to the central node (base station) will decrease the network's overall work. PCA is applied to reduce broadcasting data between sensor nodes by discovering a small set of uncorrelated, linear mixture of unusual readings. It reduces the buffer excesses at the sensor nodes or cluster heads in event-driven applications, which evade the jamming difficulty. More than a few models of WSNs, like data aggregation, target tracking, routing, and localization, have accepted PCA.

2.5.2.6 Independent Component Analysis

Independent component analysis (ICA) is more advanced than PCA, as it discovers new fundamentals for data diagrams as well as breaks down multivariate observations into additive subcomponents. ICA evaluates several data domains, like web content, social networking, digital images, business intelligence, and psychometric measurements. In numerous applications, data interpretations are time-series or a set of parallel interpretations; to differentiate these interpretations, ICA works efficiently.

2.5.3 SEMI-SUPERVISED LEARNING

The earlier sections described supervised learning dealing with labeled data and unsupervised learning dealing with unlabeled data. However, most real-world applications do not contain both. To overcome this, a new learning procedure is known as semi-supervised learning, which deals with labeled and unlabeled facts. It has two different targets: first, forecast the labels on unlabeled data in the training set and, second, expect the labels on upcoming test data sets. Based on these two targets, semi-supervised models are isolated into transductive learning, which estimates the accurate labels for the existing unlabeled data set, and inductive learning, which expects the prediction for upcoming data. Video surveillance, social media behavior, natural language processing, spam filtering, web content filtration, gene classification, and speech reorganization are some of the WSN applications best suited for semi-supervised models.

2.5.4 REINFORCEMENT LEARNING

Reinforcement learning (RL) is a special kind of learning in the ML community. When it knows a small amount of knowledge or information regarding a particular environment in the RL model, an agent or decision-maker continuously learns or adopts or collects records to take positive proceedings. It increases the model performance by shaping the most favorable consequence from the surrounding atmosphere. These kinds of operations are significantly suitable for wireless radio learning and adapted to the surrounding environment. Q-learning procedures are model-free RL models in which every agent interrelates with the environment and creates a series of inspections, like state–action–rewards. Then, a variety of techniques are utilized to explain the RL-WSN difficulties, like actor-critic (AC), Markov decision process (MDP), and policy gradient, along with deep RL (DRL). Energy harvesting, query processing, spectrum sensing, dynamic routing, and spectrum access in cognitive radio networks are some of the application of WSN interested in RL implementation.

2.5.4.1 Q-Learning

RL makes a network supervise its internal operations gracefully and takes decisions autonomously between devices with no human interface. DRL is heavily applied in large-scale power control, resource management, efficient task scheduling, channel access, interference coordination, and the like. The predictable RL is apt to make decisions with preprocessed features or else low-dimensional data. In contrast, DRL is proficient in straightforwardly discovering their action-value plans from multi-faceted high-dimensional inputs. Therefore, the scheduling outline incorporates RL allows the scheduler to cleverly enlarge a relationship among the optimal act and the existing circumstances of the situation to reduce the power expenditure with the inconsistency of workloads.

2.6 ML APPROACHES IN INTELLIGENT WSNS

2.6.1 LOCALIZATION

Localization indicates the physical or geographical position of the sensor node. In most WSN applications, nodes are deployed without knowing the correct position; moreover, there is no connectivity or adequate infrastructure to find them in a deployed region after the deployment. On the other hand, recognition of node position or location will increase the assignment's overall performance. To hold such situations, we need ML employment, which can reconfigure, reprogram, or build clusters, whereby every cluster is trained to find sensor nodes' harmonization and determine the precision of location.

Bernas and Placzek (10) designed a hybrid model with k-means and FCM techniques, and simulation results improve the localization. Two techniques are applied on a cluster head that will train the entire node with received signal strength indicator (RSSI) values. Phoemphon et al. (11) developed a novel approach for extensive mixed network localization with a fuzzy logic-based vector particle swarm optimization. Moreover, the projected model has a lower complexity than other existing models. Kang et al. (12) improvised a graph-based localization model, including CNN and SVM. Here, the graph model utilized for leakage detection to trim down the error rate, and a combination of classifiers analyzes the signals to progress the classification truthfulness (13). Authors develop an advanced SVM technique, FCM sample reduction (FCMTSR), for location detection; here, FCMTSR is employed to reduce training overhead. Zhu and Wei (14) planned a novel fuzzy linguistic formation of mobile sensor nodes' localization based on RSSI differences and fuzzy location indicator identification for geometric relocation.

2.6.2 COVERAGE AND CONNECTIVITY

Coverage along with connectivity, the two topmost difficult concerns in intelligent WSN. Depending on the application a vast number of sensor nodes are employed for a particular assignment, the cost of nodes significantly increases we use numerous nodes. With the help of accurate coverage and efficient connectivity, the cost of the network and the number of sensor nodes will also decrease. A considerable part of WSN applications use random deployment compared to other techniques. Connectivity means no lonely sensor node in the setup; each node in the WSN transfers its records to the base node indirectly or via relay nodes. In WSNs, numerous models have been projected to explain coverage along with connecting difficulties.

Kim et al. (15) designed a simple, distributed SVM model to demonstrate global localization by transmitting little messages within the local nodes; doing this also reduces the communication cost. The authors planned a combined SVM model with a DT to estimate the link quality among sensor nodes. Moreover, the model's simulation outcome formed sufficient link-quality precision, condensed energy expenditure, and enhanced the network duration. Qin et al. (16) implemented k-means with FCM for detecting a well-built connected system. Here, k-means classify sensor data and FCM for separation of workload. A two-stage sleep scheduling model designed by Chen et al. (17) of RL for area coverage, the Q-learning technique decides on the proper nodes that coat the objective locale with an inadequate number of sensor nodes. An experimental result shows the models have favored locale coverage and boost the system's existence.

2.6.3 ROUTING

In WSNs, sensor nodes are installed randomly in harsh environments, whereby each node gathers information from the surroundings and conveys it to the cluster head or sink node. However, deciding routing paths is not so easy under the dynamic changes of sensor setup. Optimal routing will enhance the network's existence because the nodes' conation requires less processing capacity, low power supply, less memory, and low transmission bandwidth. Hence, the optimal routing path aims to increase the network life with less resource utilization. Notified paybacks of ML-based routing for intelligent WSNs are trimmed-down difficulty in routing through separating them to sub-routine issues and graph structures implied to reach real-time routing. Avoiding unnecessary paths, it selects the best path with an ML-based technique that will result in energy saving, delay-aware, low communication overhead, and prolong the lifetime of WSNs.

Gharajeh and Khanmohammadi (18) designed an original three-dimensional fuzzy-based routing on traffic possibility. Here, the number of adjacent nodes and their respective distances considers input and produces output as a routing possibility. DFRTP boosts the data-delivery ratio and diminishes the power utilization of the sensor nodes. A DL-based dynamic routing model was implemented Lee (19) with sink nodes as centralized transportation. The sink node primarily builds a record of essential routing trails, and from them, it categorizes the most acceptable route. This algorithm conquers the congestion plus packet loss, power management. Khan et al. (20) planned an SVM-based energy consumption of cluster heads. Jafarizadeh et al. (21) proposed a Naïve Bayes CH selection to maintain the energy balance for resourceful clustering. Lin et al. (22) designed an adaptive built-in routing protocol with a Bayesian approach, in every iteration routing introduced with new target node selection.

2.6.4 DATA AGGREGATION

In WSNs, gathering all the raw sensed data from sensor nodes is acknowledged as Data aggregation. Proficient data aggregation approaches improve network life span and stabilize energy utilization—numerous models implemented in WSNs with ML techniques. The reasons are as follows: ML assumes surrounding situations and makes decisions without any reprogramming. ML chooses the best node as the cluster head for data collection. This will increase the extensive composure energy of nodes.

Atoui et al. (23) proposed a novel power competent multivariate data reduction model through periodic aggregation by polynomial regression utility. It's well known that the reduction of sensed data at the node level will trim down communication overhead among node to the cluster head, the cluster head to the base station, and node to the base station. The authors utilized Euclidean measurement to determine the correlation among sensor data. Gisapn et al. (24) utilized linear regression for the investigation of nonrandom parameters while accumulating data. Edwards-Murphy et al. (25) developed a heterogeneous application for health care with DTs and for agriculture monitoring with honeybee-based models. Experimental results show significant accuracy in both techniques. Habib et al. (26) used fuzzy logic-based decision-making theory for wireless body networks for improvising biosensors' capability. Whereas Naïve Bayes models utilized for imbalance categorization in Yang et al. (27), it had better results than those of Habib et al. (26). Bertrand and Moonen (28) planned the PCA-type distributed adaptive model for discovering least and utmost eigenvalues. For the accurate reconstruction of data, Chidean et al. (29) employed a compressive projection PCA (CP-PCA). To improve forecast and compression, Wu et al. (30) designed an advanced PCA; it efficiently reduces communication, creates accurate predictions, and achieves higher compression quality.

2.6.5 Congestion Control

WSNs collected data transformed through communication channels; if suppose additional data are transmitted through a channel more than its capacity, congestion will occur. Packet collision, transmission rate, many-to-one data transmission, dynamic time variation, channel contention, and the like are reasons for congestion. Congestion changes a range of constraints of WSNs, such as communication complexity, processing power, energy consumption, storage, QoS, and packet delivery ratio (PDR). ML techniques are best suited for congestion-related issues in WSNs. And reasons are that with the ML advances, it is easy to estimate accurate traffic in a particular channel and proficiently locate the best routing path with minimum end delay among nodes to the cluster head to the base station. And it is possible to change radio transmission ranges according to dynamic transforms in the network.

Rezaee and Paandideh (31) proposed a fuzzy scheme for controlling congestion by minimizing the packet loss percentage with a well-organized queue execution. First, fuzzy logic was used to execute the queue, fine-tune congestion control, and finally recover congestion by harmonizing the channel stream. Moon et al. (32) developed an energy-competent data-gathering technique with a multi-face-agent RL technique. It shows high-quality traffic control and routing decision. An SVM-based congestion model designed by Ghoulipour et al. (33) shows that broadcast rates of every node are accustomed to dynamic modification in traffic.

2.6.6 Event Detection

Most often, sensor nodes are deployed in harsh locations to learn or acquire new knowledge. For this, they need to monitor the deployed regions and process the information locally to make a suitable decision. By analyzing reliable data, it is easy to make or detect potential events. The primary reason for applying ML to WSNs for event detection is that ML effectively detects events from complex, raw sensed data; moreover, the progress PDR is an effect of accomplishing proficient duty cycles.

Illiano and Lupu (34) designed a complicated regression replica to progress the precision of event detection. Authors employed KNN to discover highly interesting information from the collected raw data. Li et al. (35) extended the version of Illiano and Lupu's replica (34) by applying KNN with a query processing approach that digs out high-rated patterns from in-network stored information. The model progresses the query processing approach and balances the energy utilization of the individual node. Han et al. (36) projected an event-exposure technique with fuzzy and rule-based techniques to accelerate the detection method. The Bayesian probability base motion detection model detects moving objects in different sectors, and this model is efficiently utilized to detect numerous dislocated devices, objects, or nodes. Advanced RL-based sleep, active, and awake forecasting technique with duty cycle has been proposed by Ye and Zhang (37) for power-proficient WSNs. Authors in (38) developed a new distributed functional tangent DT to judge the quality of water from a pond with intelligent sensor nodes.

2.6.7 QoS

The range of services provided by the WSNs to various research domains are declared through QoS. QoS is categorized into application-based, node deployment, and network-based; bandwidth and power utilization are the parameters to be considered. It is tough to handle QoS for WSNs because it is related to data redundancy, energy balancing, multiple sinks, dynamic network, and the installed network's scalability.

Collotta et al. (39) planned a power-efficient model for data combined with a fuzzy scheme. However, the model collects first-rated information instead of collecting total WSNs data. The fuzzy judgment is exploited in each sensor node to discover the proper information before the cluster head. Ren et al. (40) employed a q-learning model to explain QoS in dynamic network circumstances; the model accomplishes optimal experimental analysis results. For consistent QoS, Razzaque et al. (41) introduced a distributed adaptive cooperative routing technique with trivial RL, which chooses nodes for communication for exploit consistency. Renold and Chandrakala (42) used a hybrid model based on multi-agents for selecting active nodes and energy-aware topology for effective connectivity and consistent topology coverage.

2.6.8 Spectrum Detection and Behavior Classification

ML-based classification techniques for intelligent WSN are mostly functional for security, image classification, spectrum sensing, and anomaly and interference detections. For example, consider a situation in which many intelligent sensor nodes are deployed as smart devices to contact the spectrum radio, and different sensing practices affect high-dimensional search issues. For this, we can apply SVM and KNN models for describing a channel's working status by classifying every feature vector into an active class or unfavorable class. Mainly, with the alteration of active devices, mobility, and sleep actions in WSNs, it is quite tricky for all well-groomed devices to track a similar set of laws and principles in multifaceted and dynamic situations. SVM, Bayesian models, and KNN have powerful data-processing capabilities and classify possible network behavior. These merged categorization tools have the talent to classify devices' performance aspects; moreover, they repeatedly found that the learning model categorized various activities, which can advance the identification capability and smartly set up communication regulations between the well-groomed devices.

2.6.9 CHANNEL ESTIMATION AND MOBILITY PREDICTION

For a single vehicular application, 100 to 1000 sensor nodes will be employed; if the accurate mobility prediction is determined with intelligent WSNs, it will lead to many enjoyable applications such as traffic control, mobility, and vehicle routing, among others. Frequency alterations and wireless channel variations in high-mobility vehicles straightforwardly influence the declaration and QoS. Regression and SVM models effectively employ channel parameter estimation in intelligent WSNs by using related features to discover potential unknown transformations. Regression, SVR, and Gaussian kernels are extensively applied to estimate device behavior prediction and real cross-layer handover optimization issues in intelligent WSN.

2.6.10 DEVICE CLUSTERING AND LOCALIZATION

Cluster head models are highly capable of explaining future clustering problems in intelligent WSNs. Generally, WSNs are mostly installed in impenetrable locations where huge spectrum access services allotment is tricky. Cluster models are utilized for grouping mixed devices into special groups according to their interests; by doing this, they avoid interference, increase power ability, trim collision, and achieve higher probability. Clustering models enthusiastically augment for every localization along with tracking in intelligent WSNs. To deal with the range-based multidevice locations and pathway finding issues, clustering models are capable of optimizing the clustering dispensation of enormous locality facts and filter out the acute position references; after that, they analyze the primary cluster center with the density of every dimension point.

2.6.11 DATA COMPRESSION AND INTERFACE FILTERING

Intelligent WSNs collect huge training samples and application services; here, PCA and ICA effectively decrease the quantity of sensed data via detecting the most favorable variables that boost the overall network performance and lessens model, communication, and computational complexities. Moreover, demission decrease approaches can split the preferred signal plus noise subspace; this notably reduces the additive resonance on the channel evaluation. In some cases, every dimension's input patches could associate with others; furthermore, a few magnitudes are also diverse with the noise and obstruction records, which degrades the system's performance if those inadequate obstruction data are not cleaned appropriately.

2.7 FROM ML TO DL

ML models produce great results when they are provided with preprocessed data, which means the raw data's high-quality patterns or features must be preprocessed and recognized by human experts, then the ML models work on them. ML models can explain real-world problems with neural networks, but they are not efficient in solving all the problems (3–6). Because a typical ML structure is built with three layers—an input layer, which takes prearranged or preprocessed data as input of the network; a hidden layer, which is also known as feature extraction or processing layer, a single hidden layer extract the valuable, handy data features; and, finally, an output layer, which drops the end result based on clustering, classification, and so on based on the applied ML task.

Nevertheless, the issue of WSNs has employed network changes dynamically, and it's hard to detect or recognize helpful features from a real-world environment. The learning scheme's standard data may be dissimilar in many ways, such as color, shape, audio, video, which play a crucial role in event justification. However, the central data learning wants input data; instances must be in a uniformed form, because that depends on these proceedings being categorized. That is why standard data must identify good patterns or features to boost the ML classification system. As discussed earlier, ML consists of only one hidden layer with enough hidden units to try to learn more or else fewer autonomous aspects from the input layer. DL contains more hidden layers between the input and the output layer. DL automatically extracts suitable representations for the external task; now the production of the lower layer is input to the upper layer (4–9). Starting with ordinary data, every layer extorts unique aspects from the input data, progressively amplifies aspects or characteristics more appropriate to decision-making, and limits inappropriate features.

2.8 DL

DL is considered a super-part of ML. DL is suitable for appending intelligence to WSNs with complex radio connection and large-scale topology with the abilities like perfect pattern reformation and handling complex raw data. DL follows the brain-type arrangement; the human brain controls authoritative, commanding data-processing potentialities. Every day, the human eye obtains various data types in different environmental conditions. The brain then extracts the high-rated characteristics from those data and, at last, decides on the observed environmental condition. In a DL procedure, computers need to learn from given experiences and, from that, construct a confident training model. During the training procedure, computers find out suitable weights among hidden neural nodes that can determine the high-quality features of the input data. Once the neural network has been correctly trained, a proper conclusion is given to reach a high reward. DL can discover network dynamics like traffic bottlenecks, hot spots, spectrum availability, interference distribution, congestion points, and so on by analyzing a tremendous amount of system parameters like signal change, noise, delay, paths, and loss rate, among others. Thus, DL can be considered for extremely multipart wireless communications with several nodes plus energetic link quality. Applying DL also boosts other network tasks as network security, QoS, sensing data compression, and quality of experience, among others. Additionally, the unsolved and challenging research issues in wireless communications are debated in depth, which positions the potential research inclinations of DL-based wireless communications.

When DL is applied to a particular application, we must consider some key points; those are how to give or format the input of a particular environment in a proper numerical layout to the DL input layer, how to reorganize or signify the DL output layer results (meaning), how to arrange an objective reward function that capable to iterative weight update in every neural layer, and, finally, the construction of DL scheme, including how many hidden layers, the formation of every layer, and the associations among layers.

Wireless networks hold complicated features like routing, communication signal characteristics, massive data sets, localization, channel superiority, higher computing complexity, queuing state of every node, path congestion state, and others. To handle complicated situations, ML has been widely implemented. On the other hand, a DL application in wireless communications has drawn lots of interest (see Table 2.1).

TABLE 2.1

Comparison of ML/DL Techniques for Intelligent WSNs

Category	Learning model/ Technique	Characteristics	Application
Supervised	Decision trees	Predicts the labels of data by iteratively the input data through learning tree	Path collection, loss rate, data aggregation, anomaly
	Random forest	Collection of trees, and each of which offers a classification.	Coverage, location, MAC protocols, data categorization
	Artificial neural network	Neuron units for perfect categorization of raw data	Intrusion/anomaly, localization, discover faults, routing, congestion control
	K-NN	Majority votes of neighboring	Data aggregation, channel estimation, faulty nodes, query-processing system
	Support Vector Machine	Separable inputs, easy to train	Spectrum sensing, congestion control, connectivity issues, anomaly/intrusion
	Bayesian network	Statistical models, independent outputs like Markov time series	Channel estimation, spectrum sensing, coverage, path selection, error detection, packet tracking

Category	Learning model/ Technique	Characteristics	Application
Unsupervised	K-means	Discover the most advantageous cluster heads	Feature extraction/selection, decision making, data collection
	Fuzzy-c-means	Cluster designed based on strength, distance, connectivity	Localization, connectivity, mobility
	Support Vector Dimension	Dimensionality reduction in the form of matrix factorization	Classification, path selection, frequency modulation, data aggregation, routing
	Principle component analysis	Orthogonal axis to maximize variance to reduce the dimension	Event-driven approaches, denoising, dimension reduction, feature extraction
	Independent component analysis	Data diagram, multivariate observation into subcomponents	Reduction, clustering data, decision-making
Semi-supervised	Transductive/ inductive learning	Extracting important feature for decision-making	Channel estimation, spectrum detection, prediction, device clustering, time-domain features
Reinforcement	Q-learning	Learning unknown state transition and rewards address complicated task efficient	Spectrum access, power management, optimal network, routing
Deep Learning	Auto-encoder	Reduce the input to output by learning specifications from input data	Spectrum sensing, channel estimation, denoising, faulty nodes
	Recurrent neural network	Input/output composed in autonomous sequences	Time/Event series, channel estimation
	Convolution neural network	Successful learning models for train multilayer neurons to reduce data processing	Modular classification, inference alignment, link evolution, security, feature extraction

The success or reasons for employing DL for wireless networks include the following:

1. The human brain has a remarkable capability to accept missed, unclear, imprecise, and blurred examples. For instance, we can imagine or recognize a person from a blurred or unclear image whereby some of the pixels are not visible or are missed. The same technique is beneficial in WSN circumstances because it is difficult to collect accurate samples owing to channel failure, node displacement, channel fading, and noised values.

2. DL can also be used for intrusion detection in large network models; intrusion detection is challenging in large-scale networks where a vast quantity of traffic passes through different filters.
3. As stated earlier, WSNs collect a large amount of data, and the complexity of ML with massive data increases continuously. With human brain arrangement, DL accepts a massive quantity of raw data with numerous protocol layers and concurrently monitors diverse data and formulates the first-class decision. DL models cooperate in a critical role with big data in the wireless community.
4. DL and DRL, analytically explore modulation models and build efficient error correction codes, optimal routing, and link quality estimation, increasing the WSN's work.
5. Based on surveillance, the human brain learns about things based on the learned results that guide our activities. In the same way, learning outcomes will direct us to an accurate network model for WSNs. This technique, acknowledged as DRL, is broadly used in the large-scale wireless community.

2.8.1 DL FOR INFERENCE ALIGNMENT

In WSNs, interference alignment (IA) is one of the hot research topics because of its better channel exploitation by allowing various transmitter and receiver duos. Various authors designed a multi-input multi-output (MIMO) scheme. IA researchers employed the preceding linear practice to assist transmission indications so that the intervention signal lies in a condensed dimensional subspace at every recipient. Here, the transmitter plus receiver's synchronization shatters the throughput constraint forced by the MIMO antennas' intervention glitch.

2.8.2 DL FOR MODULATION CLASSIFICATION

A DL-based modulation categorization is a potential research theme and has been mostly applied for detecting the modulation nature in the received radio signals. Peng et al. (43) proposed a novel technique by employing a convolution neural network (CNN). The authors utilized Alex-Net for the classification of modulation with different collection patterns.

2.8.3 DL FOR PHYSICAL CODING

DL has also been successfully applied for error correction procedures; Nachmani et al. (44) planned a belief propagation decoding model for a low-concentration parity check. Also, they employed the Tanner graph to improve the parity check process. In (45), weights are skilled with stochastic gradient descent to reduce the bit error rate. Gruber et al. (45) and Cammerer et al. (46), employed polar codes enhanced with a decoding algorithm skilled with DL. O'Shea and Hoydis (47) utilized an autoencoder for end-to-end communication methods for correcting signal categorization, modulation, and error correction. In addition, DL used for MIMO and chemical signal detection (48).

2.8.4 DL FOR LINK EVALUATION

WSNs consist of heterogeneous, complex network models and hybrid resource constraints, hence reducing complexity and choosing the best technique is a tricky task. Liu et al. (49) proposed an innovative technique that decreases the everyday power expenditure by properly setting up the entire obtainable outlines. A deep belief network has been employed for link-quality estimation by Hinton et al. (50). Based on the valuation outcome, the link relations that cannot be scheduled for a flow will be expelled from the link optimization practice. This procedure capably lessens the dilemma of link optimization.

2.8.5 DL APPLICATION IN THE DATA LINK LAYER

A considerable part of WSNs' issues of data link layers, like resource allocation, link evolution, and traffic predictions, have shown promising improvement with DL inclusion. Hence, bound computation with data size, efficient channel allocation can be sensitive issues solved with DL models.

2.9 DL FOR WIRELESS SIGNAL RECOGNITION

Signal recognition in WSNs has a particular purpose in various vital tasks like spectrum monitoring, cognitive radio frequency management, and channel allocation, which boosts the upcoming 5G and advanced IoT applications. Moreover, it dramatically impacts military and aerospace applications in situations like anti-jamming, device recognitions, signal reconnaissance, and so on. As a result, a great deal of study and valuable discovery models on signal detection have been done; moreover, sequences of valuable formats are mentioned in the following. Modulation recognition (MR), which is also acknowledged as automatic modulation classification (AMC), and wireless technology recognition (WTR), are the two essential concepts in wireless signal recognition (WSR).

With the continuous development in the fields of military surveillance in tracking and monitoring, as well as civilian applications, which desperately need advanced improvements in communication and electromagnetic environment, the potential gratitude of broadcast signals has also made significant growth plus regularly turns into development. In WSNs, wireless communication provides unique advantages from the parameters like symbol rate evolution, frequency, modulation type, bandwidth estimation, and the like for the classification of signals (64–68, 72, 73).

2.9.1 WTR

Recently, WTR has received attention from various research fields with the progression in communication technology. In general, dissimilar radio pieces of utensils make for an increasing shortage of spectrum resources. Because of this, the transmission interference is inevitable in a coexisting environment, which leads to a rejection in spectrum effectiveness—besides, the mixture of network devices provides a reason for more complexity. With advanced ML and DL, the WTR favors explicit aspects of time, frequency, and time-frequency features.

2.9.1.1 Time Domain Features

Time-related characteristics are the most precise representation for a signal with phase and amplitude. Yang et al. (51) designed a CNN-based classifier by employing time-related characteristic Quadrature (IQ) vectors and amplitude. The designed model has best suited for Wi-Fi, Bluetooth, and Zigbee. Selim et al. (52) employed both amplitude and phase dissimilarities for CNN network training. The model shows excellent recognition results on radar signals under long-term evolution (LTE) (53). CNN model deeply tuned with a deep residual base on previous works, a simulation done with various channel parameters and scales. Riyaz et al. (54) have moved one step ahead by developing a radio fingerprint model by applying the IQ vector data set to CNN. The presented method is competent in learning inbuilt signatures from different wireless transmitters, functional for discovering hardware devices.

2.9.1.2 Frequency Domain Features

Frequency domain features contain more useful information than the time domain with the help of power spectral density and center bandwidth, which are crucial in WTR. A fuzzy logic scheme was developed by Ahmad et al. (55) for the recognition of wireless Local Access Network (WLAN) and Bluetooth signals. The authors applied power spectral density to extract the bandwidth and central frequency for labeling the signals. Ahmad et al. (56) extend earlier work (55) by employing a neuro-fuzzy signal classifier (NFSC) to identify nano-NET, Atmel, and Bluetooth signals. Experimental results show the model's performance is enhanced by using wideband and narrowband data to gain real-time coexistence approaches. Rajendran et al. (57) proposed a time-domain record, which also maps into a frequency-domain sign via fast Fourier transform (FFT). The outcomes confirm that the CNN formed skills from the FFT data, which significantly enhanced the precision compared to time-domain characteristics. A combined scale FFT plus LSTM model with scattered sensor design has been developed by Schmidt et al. (58), and the experimental results show that it has useful classification in LTE, Global System for Mobile Communications, and radar and WFM signals. Authors of (59) proposed a CNN-based frequency domain data capable of categorizing IEEE 802.x signals.

2.9.1.3 Time–Frequency Domain Features

Both time and frequency domains have their individual, respective advantages. However, the combination of the time–frequency domain has some unique advantages in signal recognition. Bitar et al. (60) proposed that frequency–time signs like central bandwidth, spectral bandwidth, and temporal width are forces for NNs. Longi et al. (61) advanced CNNs to classify the IEEE 802.x protocol relative to working with the ISM band. Time–frequency power standards with a matrix form are given as input to the CNN classifier. Singh et al. (62) introduced successions of time-slice spectrum data with a pseudo-format trained with semi-supervised CNN. The end results show that the designed form executes well in device identification, even with fewer labeled records in the training procedure.

2.9.2 MR

MR has developed significantly with a mixture of a variety of techniques. MR is classified into likelihood-based (LB) and feature-based (FB). LB models are capable of gaining the most favorable performance by considering all the unknown quantities for the probability density function (PDF). LB models are mostly applied to the academic progression of MR. FB models typically extort high-rated features from the received signals and have tremendous enhancement with DL inclusion. Spectrum sensing enhances the perception capability by narrowing and deciding on a spectrum availability at a particular time and the geological position. To decide on a particular frequency band, spectrum sensing decides between two hypotheses (H0 and H1), after the deficiency and the licensed user signal's existence, correspondingly.

Dobre et al. (63) stated that the corresponding filter detector associates the established signal with a well-known replica of the licensed user sign to take full advantage of the time-honored signal-to-noise ratio (SNR). Wong and Nandi (64) evaluated the energy of the observed signal with a threshold and made a verdict on the licensed abuser signal; however, it is vulnerable to noise-power-level ambiguity. Lallo (65) features an evaluation of the recurring autocorrelation of the received signal. It discriminates licensed user signals from the intervention and noise and smoothly works in deficient SNR rules. Yu et al. (66) designed a novel pilot prototype by employing subcarriers free from the interference proposed for the orthogonal-frequency-division-multiplexing-based cognitive radio, and the channel estimation was studied. Zhou et al. (67) studied cyclo-stationarity for automatic modulation signal categorization. Liu et al. (68) stated that a secondary abuser could modify their transmitting energy and maximize interference adaptively. Hassan et al. (69) used spectrum dimension arrangements designed with diminished activities. Ali and Yangyu (70) utilized the distribution of combined interfering from cognitive radio abusers to a licensed abuser to explain the sensitivity and transmitted power requisites. Ali et al. (71) studied a complete statistical attitude of the interference of cognitive radio.

2.10 OPEN RESEARCH ISSUES

1. The burst signal is extensively employed for military applications for dynamic spectrum admission. The signal contains features like shaky start and end timings and works short in a general burst. These features possess big challenges to signal recognition, data preprocessing, and decision-making, which are essential for suitable precision in signal recognition.
2. Signal screening and inference managing under the circumstance of multi-faced signal coexistence has received attention. Recent technical advances like 5G and IoT communications need a crowded spectrum, and resources are likely to overlap with multifaced signals; advanced methods are needed to solve this overlapping.
3. However, the type of signal recognizable with prearranged or copied signals is not easy to find; signal types, in advance, and unanticipated interference happen at any moment.

4. The transmission power expenditure of sensor nodes for any model checking is high, so it needs to be optimized to diminish the transmitting power expenditure.
5. There is a need to design advanced techniques, where every local node will enclose faulty event models, then transmit condensed data consequences to the centralized node.
6. Even though numerous data aggregation and data cleaning methods were applied for trim-down data transmission, large WSNs with heterogeneous devices continuously gather many raw data and are too pricey to be modernized. Figures 2.1 and 2.2 show the articles published and algorithms proposed so far, but there is still a huge demand for advanced ML and DL techniques to enhance network throughput.

2.11 CONCLUSION

WSNs contain numerous small, low-priced sensor nodes that can sense, communicate, process, and compute; taken as a whole, it is an essential applicable model of computing and networking. Traditional WSNs do not produce effective results with continuously generating data sets. Because of the close relationship in ML and DL working styles with WSNs, researchers and industrialists found a stepping-stone for intelligent WSNs. Hence, we described various reasons for successful reasons for employing ML and DL in WSN and WSR throughout this chapter. Moreover, we described various application areas and open research issues according to WSNs' existing work.

REFERENCES

1. Yang, Helin, Xianzhong Xie, and Michel Kadoch. (2020). Machine Learning Techniques and a Case Study for Intelligent WSNs. *IEEE Network*, PP(99), 1–8.
2. Alsheikh, M. A., S. Lin, D. Niyato, and H. P. Tan. (2018). Machine Learning in Wireless Sensor Networks: Algorithms Strategies and Applications. *IEEE Communication Survey and Tutorials*, 16(4), 2014, 1996–2018.
3. Li, Xiaofan, Fangwei Dong, Sha Zhang, and Weibin Guo. (2019). A Survey on Deep Learning Techniques in Wireless Signal Recognition. *Wireless Communications and Mobile Computing*, doi:10.1155/2019/5629572.
4. Zhou, Xiangwei, Mingxuan Sun, Geoffrey Ye Li, Biing-Hwang (Fred) Juang. (2020). Intelligent Wireless Communications Enabled by Cognitive Radio and Machine Learning. *IEEE Xplore*, doi:10.1109/MWC.2020.9023916.
5. Mao, Q., F. Hu, and Q. Hao. (2018). Deep Learning for Intelligent Wireless Networks: A Comprehensive Survey. *IEEE Communications Surveys & Tutorials*, 20(4), 2595–2621, Fourth Quarter.
6. Praveen Kumar, D., Tarachand Amgoth, and Chandra Sekhara Rao Annavarapu. (2018). Machine Learning Algorithms for Wireless Sensor Networks: A Survey. *Information Fusion*, 49.
7. Ayoubi, S. (2018). Machine Learning for Cognitive Network Management. *IEEE Communications Magazine*, 56(1), 158–165, January.
8. Jiang, C. (2017). Machine Learning Paradigms for Next-Generation Wireless Networks. *IEEE Wireless Communication*, 24(2), 98–105, April.

9.   Luna, Francisco, Juan F. Valenzuela-Valdés, Sandra Sendra, and Pablo Padilla. (2018). Intelligent Wireless Sensor Network Deployment for Smart Communities. *IEEE Communications Magazine*, 56, August.

10. Bernas, M., and B. Placzek. (2015). Fully Connected Neural Networks Ensemble with Signal Strength Clustering for Indoor Localization in Wireless Sensor Networks. *International Journal of Distributed Sensor Networks*, 11(12), 1–10.

11. Phoemphon, S., C. So-In, and D. T. Niyato. (2018). A Hybrid Model Using Fuzzy Logic and an Extreme Learning Machine with Vector Particle Swarm Optimization for Wireless Sensor Network Localization. *Applied Soft Computing*, 65, 101–120.

12. Kang, J., Y. J. Park, J. Lee, S. H. Wang, and D. S. Eom. (2018). Novel Leakage Detection by Ensemble CNN-SVM and Graph-Based Localization in Water Distribution Systems. *IEEE Transactions on Industrial Electronics*, 65(5), 4279–4289.

13. Baccar, N., and R. Bouallegue. (2016). Interval Type 2 Fuzzy Localization for Wireless Sensor Networks. *EURASIP Journal on Advances in Signal Processing*, 1, 1–13.

14. Zhu, F., and J. Wei. (2016). Localization Algorithm for Large-Scale Wireless Sensor Networks Based on FCMTSR—Support Vector Machine. *International Journal of Distributed Sensor Networks*, 12(10), 1–12.

15. Kim, W., M. S. Stankovi, K. H. Johansson, and H. J. Kim. (2015). A Distributed Support Vector Machine Learning Over Wireless Sensor Networks. *IEEE Transactions on Cybernetics*, 45(11), 2599–2611.

16. Qin, J., W. Fu, H. Gao, and W. X. Zheng. (2017). Distributed k-Means Algorithm and Fuzzy c-Means Algorithm for Sensor Networks Based on Multiagent Consensus Theory. *IEEE Transactions on Cybernetics*, 47(3), 772–783.

17. Chen, H., X. Li, and F. Zhao. (2016). A Reinforcement Learning-Based Sleep Scheduling Algorithm for Desired Area Coverage in Solar-Powered Wireless Sensor Networks. *IEEE Sensors Journal*, 16(8), 2763–2774.

18. Gharajeh, M. S., and S. Khanmohammadi. (2016). DFRTP: Dynamic 3D Fuzzy Routing Based on Traffic Probability in Wireless Sensor Networks. *IET Wireless Sensor Systems*, 6, 211–219.

19. Lee, Y. (2017). Classification of Node Degree Based on Deep Learning and Routing Method Applied for Virtual Route Assignment. *Ad Hoc Networks*, 58, 70–85.

20. Khan, F., S. Memon, and S. H. Jokhio. (2016). Support Vector Machine-Based Energy Aware Routing in Wireless Sensor Networks. *Robotics and Artificial Intelligence (ICRAI), 2016 2nd International Conference on IEEE*, 1–4.

21. Jafarizadeh, V., A. Keshavarzi, and T. Derikvand. (2017). Efficient Cluster Head Selection Using Naive Bayes Classifier for Wireless Sensor Networks. *Wireless Networks*, 23(3), 779–785.

22. Liu, Z., M. Zhang, and J. Cui. (2014). An Adaptive Data Collection Algorithm Based on a Bayesian Compressed Sensing Framework. *Sensors*, 14(5), 8330–8349.

23. Atoui, I., A. Makhoul, S. Tawbe, R. Couturier, and A. Hijazi. (2016). Tree-Based Data Aggregation Approach in Periodic Sensor Networks Using Correlation Matrix and Polynomial Regression. *Computational Science and Engineering (CSE) and IEEE International Conference on Embedded and Ubiquitous Computing (EUC) and 15th International Symposium on Distributed Computing and Applications for Business Engineering (DCABES), 2016 IEEE International Conference on IEEE*, 716–723.

24. Gispan, L., A. Leshem, and Y. Be'ery. (2017). Decentralized Estimation of Regression Coefficients in Sensor Networks. *Digital Signal Processing*, 68, 16–23.

25. Edwards-Murphy, F., M. Magno, P. M. Whelan, J. OHalloran, and E. M. Popovici. (2016). WSN: Smart Beehive with Preliminary Decision Tree Analysis for Agriculture and Honey Bee Health Monitoring. *Computers and Electronics in Agriculture*, 124, 211–219.

26. Habib, C., A. Makhoul, R. Darazi, and C. Salim. (2016). Self-Adaptive Data Collection and Fusion for Health Monitoring Based on Body Sensor Networks. *IEEE Transactions on Industrial Informatics*, 12(6), 2342–2352.

27. Yang, H., S. Fong, R. Wong, and G. Sun. (2013). Optimizing Classification Decision Trees by Using Weighted Naive Bayes Predictors to Reduce the Imbalanced Class Problem in Wireless Sensor Network. *International Journal of Distributed Sensor Networks*, 9(1), 1–15.

28. Bertrand, A., and M. Moonen. (2014). Distributed Adaptive Estimation of Covariance Matrix Eigenvectors in Wireless Sensor Networks with Application to Distributed PCA. *Signal Processing*, 104, 120–135.

29. Chidean, M. I. et al. (2016). Energy Efficiency and Quality of Data Reconstruction Through Data-Coupled Clustering for Self-Organized Large-Scale WSNs. *IEEE Sensors Journal*, 16(12), 5010–5020.

30. Wu, M., L. Tan, and N. Xiong. (2016). Data Prediction, Compression, and Recovery in Clustered Wireless Sensor Networks for Environmental Monitoring Applications. *Information Sciences*, 329, 800–818.

31. Rezaee, A. A., and F. Pasandideh. (2018). A Fuzzy Congestion Control Protocol Based on Active Queue Management in Wireless Sensor Networks with Medical Applications. *Wireless Personal Communications*, 98(1), 815–842.

32. Moon, S. H., S. Park, and S. J. Han. (2017). Energy Efficient Data Collection in Sink-Centric Wireless Sensor Networks: A Cluster-Ring Approach. *Computer Communications*, 101, 12–25.

33. Gholipour, M., A. T. Haghighat, and M. R. Meybodi. (2017). Hop-by-Hop Congestion Avoidance in Wireless Sensor Networks Based on Genetic Support Vector Machine. *Neurocomputing*, 223, 63–76.

34. Illiano, V. P., and E. C. Lupu. (2015). Detecting Malicious Data Injections in Event Detection Wireless Sensor Networks. *IEEE Transactions on Network and Service Management*, 12(3), 496–510.

35. Li, Y., H. Chen, M. Lv, and Y. Li. (2017). Event-Based k-Nearest Neighbors Query Processing Over Distributed Sensory Data Using Fuzzy Sets. *Soft Computing*, 1–13.

36. Han, Y., J. Tang, Z. Zhou, M. Xiao, L. Sun, and Q. Wang. (2014). Novel Itinerary-Based KNN Query Algorithm Leveraging Grid Division Routing in Wireless Sensor Networks of Skewness Distribution. *Personal and Ubiquitous Computing*, 18(8), 1989–2001.

37. Ye, D., and M. Zhang. (2017). A Self-Adaptive Sleep/Wake-Up Scheduling Approach for Wireless Sensor Networks. *IEEE Transactions on Cybernetics*, PP(99), 1–14.

38. Chandanapalli, S. B., E. S. Reddy, and D. R. Lakshmi. (2017). DFTDT: Distributed Functional Tangent Decision Tree for Aqua Status Prediction in Wireless Sensor Networks. *International Journal of Machine Learning and Cybernetics*, 1–16.

39. Collotta, M., G. Pau, and A. V. Bobovich. (2017). A Fuzzy Data Fusion Solution to Enhance the QoS and the Energy Consumption in Wireless Sensor Networks. *Wireless Communications and Mobile Computing*, 1–10.

40. Ren, L., W. Wang, and H. Xu. (2017). A Reinforcement Learning Method for Constraint-Satisfied Services Composition. *IEEE Transactions on Services Computing*, PP(99), 1–14.

41. Razzaque, M. A., M. H. U. Ahmed, C. S. Hong, and S. Le. (2014). QoS-Aware Distributed Adaptive Cooperative Routing in Wireless Sensor Networks. *Ad Hoc Networks*, 19(Supplement C), 28–42.

42. Renold, A. P., and S. Chandrakala. (2017). MRL-SCSO: Multiagent Reinforcement Learning-Based Self-Configuration and Self-Optimization Protocol for Unattended Wireless Sensor Networks. *Wireless Personal Communications*, 96(4), 5061–5079.

43. Peng, S., H. Jiang, H. Wang, H. Alwageed, and Y. Yao. (2017). Modulation Classification Using Convolutional Neural Network Based Deep Learning Model. In *Proceedings 26th Wireless and Optical Communication Conference (WOCC 2017)*, 1–5, April.

44. Nachmani, E., E. Marciano, D. Burshtein, and Y. Beery. (2017). RNN Decoding of Linear Block Codes. *arXiv Preprint*, arXiv:1702.07560, February.
45. Gruber, T., S. Cammerer, J. Hoydis, and S. Brink. (2017). On Deep Learning-Based Channel Decoding. *Proceedings IEEE 51st Annual Conference on Information Sciences and Systems (CISS 2017)*, 1–6, March.
46. Cammerer, S., T. Gruber, J. Hoydis, and S. T. Brink. (2017). Scaling Deep Learning-Based Decoding of Polar Codes via Partitioning. *IEEE Global Communications Conference*, eprint arXiv:1702.06901, February.
47. O'Shea, T., and J. Hoydis. (2017). *An Introduction to Deep Learning for the Physical Layer*, eprint arXiv:1702.00832, July, https://arxiv.org/abs/1702.00832.
48. Farsad, N. and A. Goldsmith. (2017). *Detection Algorithms for Communication Systems Using Deep Learning*, eprint arXiv:1705.08044, July, https://arxiv.org/pdf/1705.08044.pdf.
49. Liu, L., Y. Cheng, L. Cai, S. Zhou, and Z. Niu. (2017). Deep Learning-Based Optimization in Wireless Network. *2017 IEEE International Conference on Communications (ICC 2017), Paris, France*, 21–25, May.
50. Hinton, G. E., S. Osindero, and Y. W. Teh. (2006). A Fast Learning Algorithm for Deep Belief Nets. *Neural Computation*, 18(7), 1527–1554, July.
51. Kulin, M., T. Kazaz, I. Moerman, and E. De Poorter. (2018). End-to-End Learning from Spectrum Data: A Deep Learning Approach for Wireless Signal Identification in Spectrum Monitoring Applications. *IEEE Access*, 6, 18484–18501.
52. Selim, A., F. Paisana, J. A. Arokkiam, Y. Zhang, L. Doyle, and L. A. DaSilva. (2017). Spectrum Monitoring for Radar Bands Using Deep Convolutional Neural Networks. *Proceedings of the 2017 IEEE Global Communications Conference, GLOBECOM 2017*, 1–6, December.
53. O'Shea, T. J., T. Roy, and T. C. Clancy. (2018). Over-the-Air Deep Learning Based Radio Signal Classification. *IEEE Journal of Selected Topics in Signal Processing*, 12(1), 168–179.
54. Riyaz, S., K. Sankhe, S. Ioannidis, and K. Chowdhury. (2018). Deep Learning Convolution Neural Networks for Radio Identifcation. *IEEE Communications Magazine*, 56(9), 146–152.
55. Ahmad, K., U. Meier, and H. Kwasnicka. (2010). Fuzzy Logic-Based Signal Classification with Cognitive Radios for Standard Wireless Technologies. *Proceedings of the 2010 5th International Conference on Cognitive Radio Oriented Wireless Networks and Communications, CROWNCom 2010*, 1–5, June.
56. Ahmad, K., G. Shresta, U. Meier, and H. Kwasnick. (2010). Neuro Fuzzy Signal Classifier (NFSC) for Standard Wireless Technologies. *Proceedings of the 2010 7th International Symposium on Wireless Communication Systems, ISWCS'10*, 616–620.
57. Rajendran, S., W. Meert, D. Giustiniano, V. Lenders, and S. Pollin. (2018). Deep Learning Models for Wireless Signal Classification with Distributed Low-Cost Spectrum Sensors. *IEEE Transactions on Cognitive Communications and Networking*, 4(3), 433–445.
58. Schmidt, M., D. Block, and U. Meier. (2017). Wireless Interference Identification with Convolutional Neural Networks. *Proceedings of the 15th IEEE International Conference on Industrial Informatics, INDIN 2017*, 180–185.
59. Mody, A. N., S. R. Blatt, and D. G. Mills. (2017). Recent Advances in Cognitive Communications. *IEEE Communications Magazine*, 45(10), 54–61.
60. Bitar, N., S. Muhammad, and H. H. Refai. (2017). Wireless Technology Identification Using Deep Convolution Neural Networks. *Proceedings of the 28th Annual IEEE International Symposium on Personal, Indoor and Mobile Radio Communications, PIMRC 2017*, 1–6, October.

61. Longi, K., T. Pulkkinen, and A. Klami. (2017). Semi-Supervised Convolution Neural Networks for Identifying Wi-Fi Interference Sources. *Proceedings of the Asian Conference on Machine Learning*, 391–406.

62. Singh, S. P., A. Kumar, H. Darbari, L. Singh, A. Rastogi, and S. Jain. (2017). Machine Translation Using Deep Learning: An Overview. *Proceedings of the 1st International Conference on Computer, Communications and Electronics, COMPTELIX 2017*, 162–167, July.

63. Dobre, O. A., Y. Bar-Ness, and W. Su. (2003). Higher-Order Cyclic Cumulants for High Order Modulation Classification. *Proceedings of the MILCOM 2003–2003 IEEE Military Communications Conference*, 1, 112–117.

64. Wong, M. L. D., and A. K. Nandi. (2004). Automatic Digital Modulation Recognition Using Artificial Neural Network and Genetic Algorithm. *Signal Processing*, 84(2), 351–365.

65. Lallo, P. R. U. (1999). Signal Classification by Discrete Fourier Transform. *Proceedings of the Conference on Military Communications (MILCOM'99)*, 1, 197–201.

66. Yu, Z., Y. Q. Shi, and W. Su. (2003). M-Array Frequency Shift Keying Signal Classification Based-on Discrete Fourier Transform. *Proceedings of the IEEE Military Communications Conference, MILCOM 2003*, 1167–1172.

67. Zhou, L., Z. Sun, and W. Wang. (2017). Learning to Short-Time Fourier Transform in Spectrum Sensing. *Physical Communication*, 25, 420–425.

68. Liu, Z., L. Li, H. Xu, and H. Li. (2018). A Method for Recognition and Classification for Hybrid Signals Based on Deep Convolution Neural Network. *Proceedings of the 2018 International Conference on Electronics Technology, ICET 2018*, 325–330, May.

69. Hassan, K., I. Dayoub, W. Hamouda, and M. Berbineau. (2010). Automatic Modulation Recognition Using Wavelet Transform and Neural Networks in Wireless Systems. *EURASIP Journal on Advances in Signal Processing*, 42, Article ID 532898.

70. Ali, A., and F. Yangyu. (2017). Unsupervised Feature Learning and Automatic Modulation Classification Using Deep Learning Model. *Physical Communication*, 25, 75–84.

71. Ali, A., F. Yangyu, and S. Liu. (2017). Automatic Modulation Classification of Digital Modulation Signals with Stacked Auto-Encoders. *Digital Signal Processing*, 71, 108–116.

72. Patterson, J., and A. Gibson. (2017). *Deep Learning: A Practitioner's Approach.* Sebastopol, CA: O'Reilly Media, Inc.

73. Thamilarasu, G., and S. Chawla. (2019). Towards Deep-Learning-Driven Intrusion Detection for the Internet of Things. *Sensors*, 19(9).

3 Machine Learning-Based Optimal Wi-Fi HaLow Standard for Dense IoT Networks

M. Mahesh and V.P. Harigovindan

CONTENTS

DOI: 10.1201/9781003107477-3

3.1 INTRODUCTION

The Internet of Things (IoT) has received significant popularity with the rise in the deployment of wireless devices all around the globe. The concept of IoT empowers the internet to be even more collaborative and ubiquitous. Besides facilitating easy access and interaction among wireless devices like home appliances, surveillance cameras, vehicles, and many others, the IoT promotes the advancements of various applications to provide sophisticated services to peoples, industries, and government agencies. Due to the rapid increase in IoT nodes, the allocation of wireless resources has become predominant. To enable the IoT, Sigfox and LoRaWAN are widely used to connect numerous battery-constrained IoT nodes. However, Sigfox is a proprietary and operator-based network with restrictions on the deployments, packet size, and so on (1). On the other hand, although LoRaWAN is one of the most popular technologies, it is limited in terms of scalability (2).

Indeed, neither of these two technologies is yet able to meet the requirements of the IoT industry, thus promoting the new communication standards such as IEEE 802.11ah (3). Recognized as a well-known IEEE 802.11 Wireless Local Area Network (WLAN) technology, the 802.11ah standard uses sub-1-GHz license-exempt industrial, scientific, and medical bands. It introduces various enhancements at the Medium Access Control (MAC) layer (4, 5). Few of them include hierarchical association identifiers, Traffic Indication Map (TIM) segmentation, and the Target Wake Time (TWT) mode, among others. Above all, one of the significant enhancements of IEEE 802.11ah is the Restricted Access Window (RAW) mechanism, which is introduced to mitigate the effect of collisions by the vast number of nodes. Using the RAW mechanism, the access point (AP) partitions the associated nodes and allocates each group with a restricted interval to avoid the contention of the other group of nodes. These restricted intervals are known as RAW slots. However, there are several RAW configuration issues:

- Assigning more RAW slots reduces the density of the nodes per group and thereby results in underutilization of channel resources, whereas contention increases due to the few RAW slots. Hence, RAW slots should be optimally chosen.
- Most of the grouping schemes in the literature not consider the requirements of the nodes (6, 7).
- Third, adapting the RAW slot with the traffic requirements of the IoT nodes is necessary. For example, a sensor measuring the level of the fluid in an industry or the moisture sensor in an agriculture field does not have similar sampling intervals and traffic requirements. The former has higher traffic intensity than the latter. Suppose both nodes are allocated the same RAW slot. Then, the node with high traffic intensity cannot communicate enough information in a RAW slot, and it has to wait for the next RAW slot in the consecutive RAW period, which introduces the unwanted delays. Hence, the RAW slots should be proportionally allocated according to the traffic requirements of the nodes.

As the nodes in IEEE 802.11ah are usually low-end systems with limited energy, memory, and computational resources, the optimization of the RAW mechanism analytically introduces computational overhead that degrades the AP's performance. Hence, machine learning (ML) would be a feasible solution for resolving this problem (8). In this chapter, a fuzzy logic system (FLS), a multilayer perceptron artificial neural network (MLP-ANN) and an adaptive neuro-fuzzy inference system (ANFIS) are exploited to meet the challenges in configuring the RAW mechanism. Fuzzy logic is chosen because of its decision-making (9). On the other hand, ANN is very popular for relating dependent and independent variables (10). However, the major drawback of the fuzzy logic inference system (FIS) is its dependence on the number and type of membership functions and the rules governing them (11). At the same time, the neural network (NN) suffers from a lack of a definite set of rules (12). To overcome this problem, ANFIS is used because of its exceptional performance in function approximation and input–output correlation (13). ANFIS is a hybrid approximation model that combines the learning ability of the NN and the reasoning ability of the fuzzy set theory (14, 15). Because the performance of the RAW mechanism depends on the parameters like associated devices, Modulation and Coding Schemes (MCS), RAW period, and so on, FLS, ANN, and ANFIS consider these as inputs for optimizing the RAW mechanism. Exploiting the effectiveness of these techniques, ML-based optimal Wi-Fi HaLow is designed for the performance enhancement of dense IoT networks.

The remainder of the chapter is as follows: Section 3.2 introduces the IEEE 802.11ah standard. The architecture of FLS, ANN, and ANFIS is detailed in Section 3.3. Section 3.4 resolves the challenges in configuring the RAW mechanism by using ML techniques. Finally, Section 3.5 presents the conclusions of the chapter.

3.2 IEEE 802.11AH STANDARD—WI-FI HALOW

With the evolution of various networking technologies for IoT, we have witnessed a revolution in the deployment of wireless nodes. Although the IEEE 802.11 standard has gained the most popularity, its performance is severely degraded in dense networks. Forecasting the future of IoT, the IEEE 802.11ah is introduced and marketed as Wi-Fi HaLow (16). The motivation behind Wi-Fi HaLow is to meet the requirements of IoT applications, backhaul aggregation, extended range hotspot, and cellular offloading. Furthermore, the IEEE 802.11ah standard provides various enhancements to the physical and MAC layer of the legacy IEEE 802.11 standard.

3.2.1 Physical Layer

The physical layer of IEEE 802.11ah is down-clocked ten times to adapt with the sub-1-GHz spectrum. This standard uses five different channels from 1 MHz, 2 MHz, 4 MHz, 8 MHz, and 16 MHz; among these, 1 MHz and 2 MHz are mandatory. The 1-MHz channel is utilized with the robust MCS10 for a coverage of 1000 m. The next most robust scheme is MCS0 with the 2-MHz channel.

3.2.2 MAC LAYER

At the MAC layer, IEEE 802.11ah introduces various enhancements to improve the scalability, energy efficiency, and throughput of the IoT network. It also introduces a novel channel access mechanism to mitigate the effect of collisions in the network.

3.2.3 SCALABILITY

The number of association identifiers (AIDs) is limited to between 1 and 2007 in the legacy IEEE 802.11. Thus, the AP cannot be associated with not more than 2007 nodes. To bridge this gap, IEEE 802.11ah uses hierarchical AIDs to connect up to 8192 nodes, as shown in Figure 3.1. The structure of an AID is organized as pages, blocks, subblocks, and index of the nodes. The first two bits represent the index of a page, the next five represent the index of the block, and the following bits represent block, subblocks, and index of the node.

3.2.4 MINIMIZATION OF OVERHEAD

The MAC overhead of the legacy IEEE 802.11 standard exceeds by 30% for a payload of 100 bytes (17). Hence, the IEEE 802.11ah introduces several short headers, as most of the nodes in the IoT network have smaller packet size:

- Several fields in the MAC header of the legacy standard are being made optional to reduce the size of the header. For example, instead of four address fields, two address files can be sufficient for the receiver and transmitter.
- Null data packet frames are implemented to shorten the control frames.

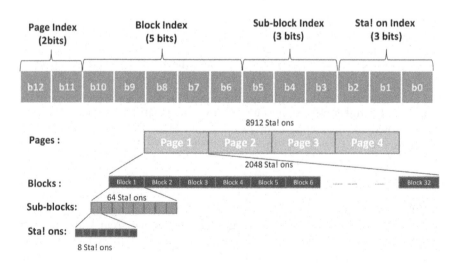

FIGURE 3.1 Hierarchical Structure of AID (18).

- The legacy beacon frames occupy significant channel time as they are transmitted at the lowest data rate. Hence, the IEEE 802.11ah uses short beacons by excluding few fields like destination address, BSSID, sequence control, and the like in the legacy MAC header.

3.2.5 Energy Management

IEEE 802.11ah conserves energy by implementing TWT mode. Based on the TIM beacons, a station periodically wakes up to check for the buffered packets destined at the AP. In case of availability of packets, the node sends the PS-Poll frame to inform the AP that it is ready for reception. However, the AP needs time to find the buffer packet and contend for the channel before sending the packet to the corresponding node. To avoid the waste of energy during this idle time, the respective node uses the TWT mode and schedules its wakeup time and enter-doze state to conserve the energy.

3.2.6 RAW Mechanism

With the aim of reducing the channel contention, IEEE 802.11ah implements the RAW mechanism, which groups the nodes and allocates a RAW slot to each group. Then, each node can select its RAW slot using Equation 3.1:

$$\xi_{slot} = (AID_N + N_{offset}) \bmod L, \tag{3.1}$$

where N_{offset} is used for fairness, and L is the configured RAW slots allocated by the AP (19). For example, a node with $AID = 5$, $N_{offset} = 0$, and $L = 5$ chooses 0th RAW slot ($\xi_{slot} = 5 \bmod 5 = 0$). Every node wakes up at target beacon transmission time and listens to the beacon frame signaled by the AP. The beacon interval comprises at least a RAW period followed by a contention-access period (CAP). The AP configures the RAW mechanism and broadcasts the RAW Parameters Set Information Elements (RPS-IE) using the beacon frame. The RPS-IE consists of the AIDs of the nodes, the duration of the RAW slot, and so on.

The RAW mechanism can be generic RAW or a triggered RAW mechanism. The former is used as the default channel access mechanism. Here, each RAW slot is used by at least one node in a dense network, whereas the RAW slots remain unused in the less dense network, as shown in Figure 3.2a, because of the lack of contending nodes. The latter is enabled by configuring the RPS-IE. In the triggered RAW mechanism, all the nodes contend in the first RAW period to reserve a slot. The AP informs the reservation of the RAW slot with the resource allocation (RA) frame. Figure 3.2b illustrates an example of a triggered RAW mechanism. Here, Group 1 and Group 3 nodes communicate with the PS-Poll frame to reserve a RAW slot in the next coming RAW period whereas Slot 2 and Slot 4 are not utilized. The AP sets the RA bit in RPS-IE and broadcasts through the beacon frames. Thus, Group 1 and Group 3 are allocated with slots in the consecutive RAW period, which improves the efficiency

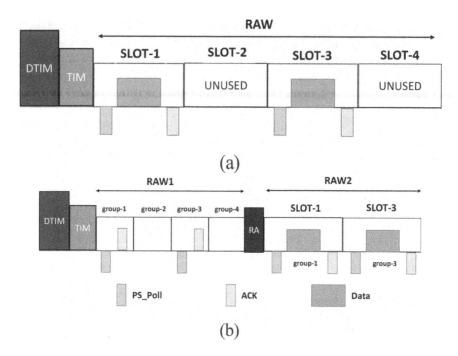

FIGURE 3.2   (a) Generic RAW, (b) Triggered RAW.

of the RAW by avoiding the unused RAW slots. However, in the case of a dense IoT network, the triggered RAW mechanism results in overhead due to the transmission of multiple RA frames and PS-Poll frames for slot reservation, which leads to the degradation of network performance.

3.3 OVERVIEW OF ML TECHNIQUES

This section introduces the relevant ML techniques used to resolve the configuration issues in the RAW mechanism.

3.3.1 FLS

Different from solving mathematical equations, fuzzy logic makes decisions based on the set of predefined rules that maps the input and output variables (20–22). The set $\Re$ in fuzzy logic is defined by membership function with linguistic terms and is given by Equation 3.2:

$$\Re = \{(x, \varphi_{\Re}(x))\ \ x \in X\},\tag{3.2}$$

where $\varphi_{\Re}(x)$ is the membership function. The membership function may be a triangular, trapezoidal, Gaussian, or any other function (8). Figure 3.3 illustrates the basic architecture of the FLS that mainly consists of the following three steps:

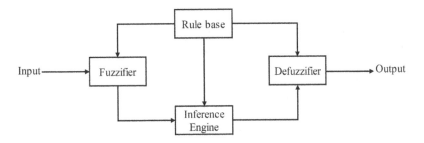

FIGURE 3.3 Fuzzy Logic Architecture.

- Fuzzification—map the input/output variable to the fuzzy membership functions
- Fuzzy inference process—derive the fuzzy output based on the fuzzy rules framed on the membership functions
- Defuzzification—calculate the output using the center of gravity (CoG) method

3.3.1.1 Fuzzification

The process of fuzzification maps the input with its fuzzy set. For example, a trapezoidal membership function $\varphi_{\Re}(x;v_1,v_2,v_3,v_4)$, which has computational efficiency and less complexity, is given by Equation 3.3:

$$\varphi_{\Re}(x) = \begin{cases} 0; & \text{for } x \leq v_1 \\ \dfrac{x-v_1}{v_2-v_1}; & \text{for } v_1 \leq x \leq v_2 \\ 1; & \text{for } v_2 \leq x \leq v_3, \\ \dfrac{v_4-x}{v_4-v_3}; & \text{for } v_3 \leq x \leq v_4 \\ 0; & \text{for } x \geq v_4 \end{cases} \tag{3.3}$$

where v_1,v_2,v_3,v_4 are the antecedent parameters that define the shape of the trapezoidal membership function. Therefore, the fuzzy universal set for the input X is given by Equation 3.4:

$$\varphi_{\Re} = \{\varphi_{Less}, \varphi_{Medium}, \varphi_{High}\}. \tag{3.4}$$

Because each input has several membership functions, the fuzzy input vector is the array of the degree of membership function of that particular fuzzy set. Therefore, the fuzzy input vector is given by Equation 3.5:

$$\bar{X} = \left\{ \frac{\varphi_{Less}(x)}{Less}, \frac{\varphi_{Medium}(x)}{Medium}, \frac{\varphi_{High}(x)}{High} \right\}. \tag{3.5}$$

3.3.1.2 Fuzzy Rules

Fuzzy rules compute the degree of membership functions. Let us consider X as input and Y as output variable with the same fuzzy universal set according to Equation 4. Then, the fuzzy rule is described as

Rule: IF x_i is A_i THEN y_i is C_i,

where A_i and C_i are the elements in the fuzzy set.

To describe the evaluation of fuzzy rules, two inputs, X_1, X_2, and one output variable, Y, are considered that have a similar fuzzy universal set according to Equation 3.4. Therefore, the fuzzy input vector for the inputs X_1 and X_2 are given by Equations 3.6 and 3.7:

$$\bar{X}_1 = \left\{ \frac{\varphi_{Less}(x_1)}{Less}, \frac{\varphi_{Medium}(x_1)}{Medium}, \frac{\varphi_{High}(x_1)}{High} \right\} = \{\bar{x}_1^1, \bar{x}_1^2, \bar{x}_1^3\}, \tag{3.6}$$

$$\bar{X}_2 = \left\{ \frac{\varphi_{Less}(x_2)}{Less}, \frac{\varphi_{Medium}(x_2)}{Medium}, \frac{\varphi_{High}(x_2)}{High} \right\} = \{\bar{x}_2^1, \bar{x}_2^2, \bar{x}_2^3\}. \tag{3.7}$$

Let us consider an example of a fuzzy rule that states the relation between X_1, X_2, and Y.

$Rule_{i,j}$: IF $\bar{x}_i^1$ AND $\bar{x}_j^1$ THEN $y = C_{ij}$.

Therefore, the strength of the rule is evaluated as

$$w_n = \bar{x}_i^1 \quad AND \quad \bar{x}_j^2. \tag{3.8}$$

Here, the AND operation is evaluated as $w_n = \bar{x}_i^1 \bar{x}_j^2$ or $w_n = min\{\bar{x}_i^1, \bar{x}_j^2\}$ based on the fuzzy set theory.

3.3.1.3 Defuzzification

FLS estimates the output using the COG method according to Equation 3.9:

$$ouput = \frac{\sum_i w_i C_i}{\sum_n w_i}. \tag{3.9}$$

3.3.2 ANNs

MLP-ANNs are used to model complex industrial problems by generalizing the relation among the system input and output parameters (10). The feed-forward ANN is most popular ANN architecture that learns the correlation among the inputs and

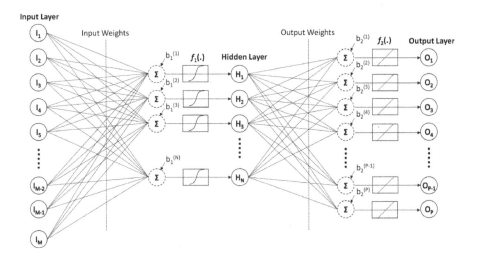

FIGURE 3.4 Architecture of an ANN (18).

outputs. As shown in Figure 3.4, the architecture of ANN has input, hidden, and output layers with at least one artificial neuron. All the neurons are programmed to evaluate the degree of association between them through appropriate weights and bias. The inputs are fed straight by layer to the output and the relation between them is given by

$$\mathbf{H} = f(\mathbf{I}_w.\mathbf{I} + \mathbf{b1}).$$

$$\mathbf{O} = f(\mathbf{O}_w.\mathbf{H} + \mathbf{b2}). \tag{3.10}$$

where I_w and O_w are given by

$$\mathbf{I}_w = \begin{bmatrix} w_{11} & w_{12} & \cdots & w_{1M} \\ w_{21} & w_{22} & \cdots & w_{2M} \\ \cdots & \cdots & \cdots & \cdots \\ w_{N1} & w_{N2} & \cdots & w_{NM} \end{bmatrix}, \mathbf{O}_w = \begin{bmatrix} w_{11} & w_{12} & \cdots & w_{1P} \\ w_{21} & w_{22} & \cdots & w_{2P} \\ \cdots & \cdots & \cdots & \cdots \\ w_{N1} & w_{N2} & \cdots & w_{NP} \end{bmatrix}. \tag{3.11}$$

The mean square error (MSE) between the current output and desired output value that is given by

$$MSE = \frac{1}{k}\sum_{i=1}^{k}(Ouput_{desired,i} - Output_{current,i})^2, \tag{3.12}$$

where k is the number of iterations. The MSE is minimized by updating the weights and biases using the back propagation of error.

3.3.3 ANFIS

ANFIS is a hybrid approximation model that uses the salient features of fuzzy logic and neural networks (15). ANFIS combines the fuzzy set theory with ANN's input–output mapping (23). ANFIS takes the essential information to tune the FIS using a back-propagation algorithm adaptively. The back-propagation algorithm minimizes the error between the actual value and the predicted value by back-propagating the error and adaptively tuning the parameters of FIS. The basic architecture of the ANFIS is shown in Figure 3.5. The ANFIS architecture has five layers in which each circle is a fixed node and each square is an adaptive node; that is, the parameters at these nodes change adaptively.

For x^{th} input to ANFIS, the first layer consists of at least one membership function (mf) defined as $\varphi_{\Re}(x;v_1,v_2,v_3,v_4)$. The first layer takes the inputs from the data fed to the NN and adaptively tunes the parameters (v_1,v_2,v_3,v_4) of the mf corresponding to each input during the training phase. Therefore, the output of this first layer $O_{1,i}$ corresponding to the jth input variable is given by

$$O_{1,i} = \varphi_{mf_{j,i}}(x_j). \tag{3.13}$$

Each node in the second layer represents a rule. Hence, this layer forms a rule base. The ith output of this layer is given by

$$O_{2,i} = r_i = \varphi_{mf_{1,i}}(x_1) \times \varphi_{mf_{2,i}}(x_2) \ldots \times \varphi_{mf_{j,i}}(x_j). \tag{3.14}$$

The third layer performs the normalization on the fuzzy rule base and gives the following output:

$$O_{3,i} = \overline{r}_i = \frac{r_i}{\displaystyle\sum_i r_i}. \tag{3.15}$$

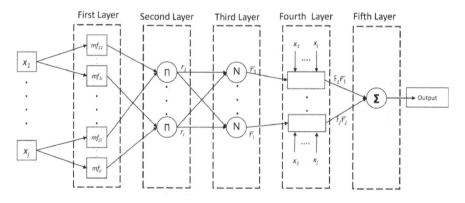

FIGURE 3.5   Architecture of ANFIS.

The fourth layer performs the defuzzification, and the output of this layer is given by

$$O_{4,i} = \bar{r}_i \mathcal{F}_i,$$ (3.16)

where $\mathcal{F}_i$ is the resulting function.

The fifth layer gives the crisp output for the corresponding inputs. The output of this layer is the summation of all the inputs and the converted crisp value:

$$O_{5,i} = \sum \bar{r}_i \mathcal{F}_i = \frac{\sum\limits_i^{max} r_i \mathcal{F}_i}{\sum\limits_i r_i}.$$ (3.17)

Finally, the crisp output is back-propagated to the NN to reduce the error (given by Equation 3.12) between the actual output and the predicted output by adaptively changing the antecedent parameters (v_1, v_2, v_3, v_4) of each membership function.

3.4 OPTIMIZATION OF THE RAW MECHANISM USING ML TECHNIQUES

This section employs the ML techniques presented in Section 3.2.3 for resolving the configuring challenges in the RAW mechanism. Mainly the following section addresses the problem of finding the optimal number of RAW slots (ORs), the grouping scheme, and the allocation of RAW slots based on the transmission requirements of the nodes.

3.4.1 ESTIMATING ORs

In this subsection, the ORs for optimizing the RAW mechanism is estimated by using fuzzy logic, ANN, and ANFIS.

3.4.1.1 Fuzzy Logic–Based Optimization Framework

In this subsection, fuzzy logic is exploited to find the ORs. The basic architecture of FLS is discussed in Section 3.3.1. The block diagram of the FLS employed in this section is shown in Figure 3.6 (24). The following are the inputs to the FLS to find the ORs: (a) associated devices, N; (2) the probability of collisions as per the model presented in Mahesh and Harigovindan (24); and (3) the MCS. The inputs N, collision probability (P_c) and MCS are defined by four, four, and nine mfs given in Equation 3.3, respectively. One hundred forty-four fuzzy rules are defined to assess the ORs. All the fuzzy variables, like mf. and fuzzy rules are intuitively considered. An example of the rule base is shown in the following discussion.

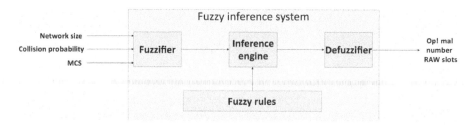

FIGURE 3.6   Block Diagram of an FLS for Finding ORs (24).

Then Equation 3.9 is used to compute the output ORs from the FLS:

$$IF\ N = LOW, and\ Pc = LOW, and\ MCS = MCI\ THEN_{\ L_{opt}=Opt1}.$$

3.4.1.2 Optimization of RAW Using ANN

This section employs ANN to establish a relationship between the associated devices, MCS, RAW period (T_{RAW}), and the ORs and generalize it to the other inputs. The data set is generated by computing the ORs analytically using the model presented in Mahesh and Harigovindan (18), and the ANN is trained with the generated data set.

3.4.1.3 ANFIS-Based Optimization Framework

In this section, the ORs are estimated by using ANFIS. The process of obtaining ORs is as follows,

Using MATLAB's gamultiobj function, the model presented in Mahesh and Harigovindan (24) is analytically solved for various network sizes and MCS to obtain the ORs:

1. The generated database is used to train the ANFIS.
2. Having been trained, the ANFIS generalizes the output for any given set of inputs, $x \in \{N, MCS\}$.

Figure 3.5 shows the architecture of the ANFIS. ANFIS forms the rule-base from the given set of inputs and outputs. An example of Takagi-Sugeno's rules governing the input and output (25) is given by

Rule 1: IF N is $mf_{1,1}$ and MCS is $mf_{2,1}$ THEN $\mathcal{F}_1 = a_1 N + b_1 MCS + c_1$

Rule 2: IF N is $mf_{1,2}$ and MCS is $mf_{2,2}$ THEN $\mathcal{F}_2 = a_2 N + b_2 MCS + c_2$,

where $mf_{1,1}$, $mf_{1,2}$ and $mf_{2,1}$, $mf_{2,2}$ are the linguistic labels of inputs N and MCS, respectively; $\mathcal{F}$ is the resulting function; and a, b, and c are the consequent parameters. Because of the less computational complexity, the trapezoidal function is used as a mf in this application. For a given set of inputs $x \in \{N, MCS\}$, the mf is given by Equation 3.3. For the inputs N and MCS, three mfs $(\varphi_{mf_{1,1}}, \varphi_{mf_{1,2}}, \varphi_{mf_{1,3}})$ and nine mfs $(\varphi_{mf_{2,1}}, \varphi_{mf_{2,2}}, \ldots, \varphi_{mf_{2,9}})$ are used, respectively. The first layer of ANFIS estimates the degree of membership function of each input. Therefore, the ith output of the first layer $O_{1,i}$ is given by

$$O_{1,i} = \varphi_{mf_{1,j}}(N) \ or \ O_{1,i} = \varphi_{mf_{2,k}}(MCS); \ j \in [1,3], k \in [1,9]. \tag{3.18}$$

Each node in the second layer represents a rule. Hence, this layer forms the rule base. The ith output of the second layer is given by

$$O_{2,i} = r_i = \varphi_{mf_{1,j}}(N) \times \varphi_{mf_{2,k}}(MCS); \ j \in [1,3], k \in [1,9]. \tag{3.19}$$

The third layer performs the normalization on the fuzzy rule base and gives the following output:

$$O_{3,i} = \bar{r}_i = \frac{r_i}{\sum\limits_i r_i}. \tag{3.20}$$

The fourth layer performs the defuzzification, and the output of this layer is given by

$$O_{4,i} = \bar{r}_i \mathcal{F}_i. \tag{3.21}$$

The fifth layer gives the K for the corresponding inputs. The output of this layer is the summation of all the inputs from the fourth layer and is given by

$$O_{5,i} = L_{opt} = \sum \bar{r}_i \mathcal{F}_i = \frac{\sum\limits_i^{max} r_i \mathcal{F}_i}{\sum\limits_i r_i}. \tag{3.22}$$

Finally, the crisp output, that is, K, is back-propagated to the NN to reduce the MSE between the actual output and the predicted output. The MSE is calculated according to Equation 3.12. The ANFIS is trained by using the hybrid algorithm to form the optimal rule base according to the input–output relation. In the forward path, the hybrid algorithm fixes $v_1, v_2, v_3,$ and v_4 and trains the consequent parameters (a, b, c) by using least-squares error method, whereas a, b, and c are fixed in a backward path and antecedent parameters (v_1, v_2, v_3, v_4) are trained by using gradient descent method (26). This process is iterated until the MSE meets a specified value or the number of iterations is reached.

3.4.2 Grouping Schemes and Adaptive Allocation of RAW Slots

In this subsection, fuzzy C-means clustering (FCM) and self-organizing maps (SOMs) are exploited to group the nodes according to their channel access requirements. Then the RAW slot is systematically formulated as per the traffic criteria of the nodes. Having associated with the AP, every node sends its transmission requirements θ_n to the AP (24).

Algorithm 1 **FCM & SOM based grouping (24)**

1. **Sorting θ_n:**

2. **for** each node **do**

3. $A(n) = \theta_n;$

4. **end for**

5. _____

6. **FCM-based grouping:**

7. $i = 0$

8. $U\,;\backslash\backslash\ ORs \times N$ matrix

9. $z = 2\,;$

10. **while** $|U^{(i+1)} - U^{(i)}|\leq 0.01$ **do**

11. $C_k = \dfrac{\sum\limits_{i=1}^{n}(u_{ki})^z d_i}{\sum\limits_{i=1}^{n}(u_{ki})^z}, k \in [1, ORs]\;;\backslash\backslash\ \text{Estimate the k-centroids}$

12. $u_{ki} = \sum\limits_{j=1}^{k}\left[(\dfrac{d_i - C_k}{d_j - C_k})^{\frac{2}{z-1}}\right]^{-1}$

13. $i = i+1$

14. **end while**

15. Group each row of the matrix U;

16. _____

17. **SOM-based grouping**

18. $epoch = 100\;;\backslash\backslash\ \text{Number of iterations}$

19. $\alpha = 0.5\;;\backslash\backslash\ \text{Learning rate}$

20. $D = [\,]\;;\backslash\backslash\ \text{Null matrix}$

21. Randomly initialize $i \times j$ weight matrix (W) i.e., $w_{ij} = rand(i, j)$, $i = 1, j = K\;;$

22. **for** each epoch **do**

23. **for** each node **do**

24. **for** each group **do**

25. $d_j = (A(n) - w_{1j})^2, j \in [1, K]$;

26. $D = [D \; d_j]$;

27. **end for**

28. \\ V is minimum value of D

29. \\ I is the index of minimum value in D

30. $[V \; I] = min(D)$;

31. Update the Ith column of w_{ij} ;

32. $w_{ij}(new) = w_{ij}(old) + \alpha(T(n) - w_{ij}(old))$;

33. $\alpha = 0.5\alpha$;

34. **end for**

35. **end for**

36. AP forms K groups in which each node joins a group with minimum weight.

3.4.2.1 FCM- and SOM-Based Grouping and Adaptive Allocation of RAW Slots

The AP uses either FCM or SOMs to classify the nodes according to their transmission requirements into K groups according to Algorithm 1. The readers are encouraged to refer Bezdek et al. (27) and Kohonen (28) for more details about FCM and SOMs, respectively. Then, the RAW slot is computed by

$$T_{slot,k} = T_{RAW} \times \begin{cases} \dfrac{C_k}{\sum\limits_{i=1}^{K} C_i}, & using \; FCM, \\ \dfrac{d_k}{\sum\limits_{i=1}^{K} d_i}, & using \; SoM. \end{cases} \qquad (3.23)$$

3.4.3 COMPARISON AMONG THE PRESENTED ML TECHNIQUES

This section compares the performance of FLS, ANN, and ANFIS techniques in optimizing the RAW mechanism. The parameters used for comparison are considered from Mahesh and Harigovindan (18, 24). Figure 3.7 presents the ORs by varying the network size and compares them with the presented ML techniques. It is clear from the figure that the analytical value of ORs is almost the same as the value of ORs found using either FLS, ANN, or ANFIS. Hence, ML techniques could be a feasible solution to optimize the RAW mechanism.

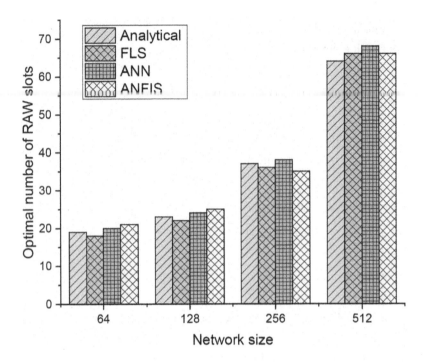

FIGURE 3.7   Comparison between ORs Using ML Techniques (18, 24).

3.5 CONCLUSION

In this chapter, the applications of ML technologies like FLS, ANN and ANFIS were presented for optimizing the RAW mechanism. Due to the computational overhead in solving the analytical equation, ML is observed to be an optimal choice to improve network performance. In this chapter, the basic architectures of the ML techniques are initially discussed. Then these techniques are applied to estimate the ORs, which play a major role in maximizing network performance. Furthermore, two simple yet efficient algorithms are presented to classify the network and allocate the RAW slots according to the transmission criteria of the nodes.

REFERENCES

1. Goursaud, C., and J. M. Gorce. (2015). Dedicated Networks for IoT: PHY/MAC State of the Art and Challenges. *EAI Endorsed Transactions on Internet of Things*, October, doi:10.4108/eai.26-10-2015.150597.
2. Adelantado, F., X. Vilajosana, P. Tuset-Peiro, B. Martinez, J. Melia-Segui, and T. Watteyne. (2017). Understanding the Limits of Lorawan. *IEEE Communications Magazine*, 55(9), 34–40.
3. IEEE Standard for Information Technology-Telecommunications and Information Exchange Between Systems Local and Metropolitan Area Networks-Specific Requirements Part 11: Wireless LAN Medium Access Control (MAC) and Physical Layer (PHY) Specifications Amendment 4: Enhancements for Transit Links Within Bridged

Networks. (2018). *IEEE Std 802.11ak-2018 (Amendment to IEEE Std 802.11(TM)-2016 as Amended by IEEE Std 802.11ai(TM)-2016, IEEE Std 802.11ah(TM)-2016, and IEEE Std 802.11aj(TM)-2018)*, 1–97, June.

4. Park, M. (2015). IEEE 802.11ah: Sub-1-GHz License-Exempt Operation for the Internet of Things. *IEEE Communications Magazine*, 53, 145–151, September.

5. Wang, H., and A. O. Fapojuwo. (2017). A Survey of Enabling Technologies of Low Power and Long Range Machine-to-Machine Communications. *IEEE Communications Surveys Tutorials*, 19, 2621–2639, Fourth quarter.

6. Gopinath, A. J., and B. Nithya. (2018). Mathematical and Simulation Analysis of Contention Resolution Mechanism for IEEE 802.11ah Networks. *Computer Communications*, 124, 87–100.

7. Kim, Y., G. Hwang, J. Um, S. Yoo, H. Jung, and S. Park. (2016). Throughput Performance Optimization of Super Dense Wireless Networks with the Renewal Access Protocol. *IEEE Transactions on Wireless Communications*, 15, 3440–3452, May.

8. Yousaf, R., R. Ahmad, W. Ahmed, and A. Haseeb. (2017). Fuzzy Power Allocation for Opportunistic Relay in Energy Harvesting Wireless Sensor Networks. *IEEE Access*, 5, 17165–17176.

9. Zadeh, L. A. (1994). Soft Computing and Fuzzy Logic. *IEEE Software*, 11, 48–56, November.

10. Hornik, K., M. Stinchcombe, and H. White. (1989). Multilayer Feedforward Networks Are Universal Approximators. *Neural Networks*, 2(5), 359–366.

11. Altin, N., and Šbrahim Sefa. (2012). dSPACE Based Adaptive Neuro-Fuzzy Controller of Grid Interactive Inverter. *Energy Conversion and Management*, 56, 130–139.

12. Kabir, Golam, and M. A. A. Hasin. (2013). Comparative Analysis of Artificial Neural Networks and Neuro-Fuzzy Models for Multicriteria Demand Forecasting. *International Journal of Fuzzy System Applications*, 3(1), 1–24.

13. Volosencu, C., and D. I. Curiac. (2013). Eÿciency Improvement in Multi-Sensor Wireless Network Based Estimation Algorithms for Distributed Parameter Systems with Application at the Heat Transfer. *EURASIP Journal on Advances in Signal Processing*, 4, January.

14. Baccar, N., M. Jridi, and R. Bouallegue. (2017). Adaptive Neuro-Fuzzy Location Indicator in Wireless Sensor Networks. *Wireless Personal Communications*, 97, 3165–3181, November.

15. Kumar, S., N. Lal, and V. K. Chaurasiya. (2018). A Forwarding Strategy Based on Anfis in Internet-of-Things-Oriented Wireless Sensor Network (WSN) Using a Novel Fuzzy-Based Cluster Head Protocol. *Annals of Telecommunications*, 73, 627–638, October.

16. Baños-Gonzalez, V., M. S. Afaqui, E. Lopez-Aguilera, and E. Garcia-Villegas. (2016). IEEE 802.11ah: A Technology to Face the IOT Challenge. *Sensors*, 16(11).

17. Zanella, A., N. Bui, A. Castellani, L. Vangelista, and M. Zorzi. (2014). Internet of Things for Smart Cities. *IEEE Internet of Things Journal*, 1, 22–32, February.

18. Mahesh, M., and V. P. Harigovindan. (2020). ANN-Based Optimization Framework for Performance Enhancement of Restricted Access Window Mechanism in Dense IoT Networks. *S̄adhan̄a*, 45, 52, February.

19. Mahesh, M., and V. P. Harigovindan. (2018). Throughput and Energy Efficiency Analysis of the IEEE 802.11ah Restricted Access Window Mechanism. In *Smart and Innovative Trends in Next Generation Computing Technologies*, Singapore: Springer, 227–237.

20. Mitra, S., and Pal, S. K. (1996). Fuzzy Self-Organization, Inferencing, and Rule Generation. *IEEE Transactions on Systems, Man, and Cybernetics—Part A: Systems and Humans*, 26, 608–620, September.

21. Lee, C. C. (1990). Fuzzy Logic in Control Systems: Fuzzy Logic Controller I. *IEEE Transac-tions on Systems, Man, and Cybernetics*, 20, 404–418, March.

22. Lee, Jihong. (1993). On Methods for Improving Performance of Pi-Type Fuzzy Logic Controllers. *IEEE Transactions on Fuzzy Systems*, 1, 298–301, November.
23. Liu, H., J. Zhou, and S. Wang. (2008). Application of Anfis Neural Network for Wire Network Signal Prediction. *2008 IEEE International Symposium on Knowledge Acquisition and Modeling Workshop*, 453–456, December.
24. Mahesh, M., and V. P. Harigovindan. (2019). Fuzzy Based Optimal and Trayc Aware Restricted Access Window Mechanism for Dense IoT Networks. *Journal of Intelligent & Fuzzy Systems*, 37, 7851–7864, June.
25. Baccar, N., M. Jridi, and R. Bouallegue. (2017). Adaptive Neuro-Fuzzy Location Indicator in Wireless Sensor Networks. *Wireless Personal Communications*, 97, 3165–3181, November.
26. Petkovi¢, D., N. D. Pavlovi¢, Žarko ‡ojbaši¢, and N. T. Pavlovi¢. (2013). Adaptive Neuro Fuzzy Estimation of Underactuated Robotic Gripper Contact Forces. *Expert Systems with Applications*, 40(1), 281–286.
27. Bezdek, J. C., R. Ehrlich, and W. Full. (1984). Fcm: The Fuzzy c-Means Clustering Algorithm. *Computers & Geosciences*, 10(2), 191–203.
28. Kohonen, T. (1990). The Self-Organizing Map. *Proceedings of the IEEE*, 78(9), 1464–1480, 1990.

4 Energy Efficiency Optimization in Clustered Wireless Sensor Networks via Machine Learning Algorithms

T. Sudarson Rama Perumal, V. Muthumanikandan, and S. Mohanalakshmi

CONTENTS

4.1 INTRODUCTION

A wireless sensor network (WSN) is nothing but a collection of many small, less expensive, self-ruling sensor nodes that require reduced power. They collect information from the global surroundings, group it, and then transfer it to a base station for further processing. With different types of sensors, the sensor nodes are used to process the data. Most of the recent research has focus on developing solutions for energy-efficient deployment, routing, and WSN management. In practice, WSN originators face a common problem linked to information clumping, information dependability, nodule bunching, energy-saving routing, events arrangement, burden exposure, and security (1).

DOI: 10.1201/9781003107477-4

Machine learning (ML) nowadays develops more algorithms that are feasible and forceful. From many application areas, voice recognition, image processing, anonymous detection, and advertisement network ML algorithms have been commonly used in classification, regression, and density estimation in recent years. The algorithms come from a wide variety of different fields, spanning statistics, mathematics, and computer science, among others (4).

The binary traditional meanings of ML include the following:

1. The development of a workstation-aided system is used to provide the solution for knowledge acquisition and to improve their performance (2, 5).
2. The machine performance should be fitted by identifying and explaining the consistencies and identifying the training data patterns (3).

In the last few years, WSNs have improved with the advancement in ML methods (7). A brief study about ML algorithms applied to sensor networks helps to improve their performance and decision-making capacities (10). The three popular ML algorithms used in all the communication layers in the sensor networks are reinforcement learning, convolutionary neural networks, and decision tree algorithms (11).

An outlier identification method has been developed and is used for some of the ML algorithms (12). There are discussions about computational intelligence methods for tackling challenges in wireless sensor networks (13). Convolutional neural networks, fuzzy systems, and evolutionary algorithms are the building blocks of ML (14, 15).

These initial reviews usually focused on the underlying learning, neural networks, and decision trees that were general in both model and strength training (8). In this chapter, we provide the many changes in the essential, most recent ML algorithms to contrast their strengths and weaknesses. Specifically, we also provide an all-inclusive overview that roughly groups these new approaches into supervised, unregulated methods of reinforcement learning. Our research argues that machine erudition procedures build on their goal WSN checks to increase the acceptance of existing WSN bids for ML clarifications.

4.2 BRIEF STARTER TO ML IN WSNS

WSN is typically characterized by sensor network designers as a group of gears and procedures used to establish forecast models (16). Specialists in ML know it as an caustic field with very large tunes and designs. ML models, applied to recurrent requests by WSNs, have tremendous elasticity advantages (19). In the context of WSNs, offerings include some hypothetical ethics and practices for implementing device learning.

The model's planned design will recognize current algorithms for machine learning. There are three categories of ML algorithms such as supervised, unsupervised, and reinforcement learning (16). ML algorithms include a named training data set in the first group. This collection constructs a model that therefore provides the relationship between the already provided input and the resultant output with respect to the parameters. Unsupervised learning algorithms, as opposed to supervised learning, are not equipped with names. An unsupervised kernel function classifies the data into clusters by means of comparing the similar features, if any, of the input data. The next category (the third) focuses on improving learning, whereby the agent learns by interacting with the situation. Hybrid algorithms focus on the inherent advantages of the main phases mentioned while reducing the vulnerabilities (17).

The examples provided clarify WSNs' progression in employing ML. Such specifics are omitted in Sections 4.3 and 4.4. Refer to Abu-Mostrafa et al. (16), and Kulkarni et al. (18) for further references about in-depth discussions of ML philosophy.

4.2.1 SUPERVISED LEARNING

In this learning, a considered exercise of fixed remains is used to form the structure model. This model is used to indicate the academic relationship among input, output, and device limitations. The big supervised learning procedures in the WSN sense continue to be debated. WSN uses these algorithms to explore the positioning and targeting objects (e.g., (19–21)). Event Detection and Query Dispensation (e.g., (22–25)), Media Access Prevention (e.g., (26–28)), Security and Intrusion Detection (e.g., (29–32)) and Quality of Service (QoS), Data Integrity and Detection of Defects (QoS) (e.g., (33–35)) for the targeting objects in WSN.

k-nearest neighbour (KNN): In Figure 4.1 this intelligence testing method categorizes an information model created from the markers of the nearby information model. Then, using a genuine solution of corresponding sensors within similar diameter limits, for example, the missing interpretations of the sensor node are predicted. To evaluate the closed set of nodes, there are a number of functions. The simplest solution is obtained by using the Euclidean distance algorithm to find the distance between the various sensors. The nearest neighbor, k, does not require large computing power/time. The factor of that function is calculated with respect to regional

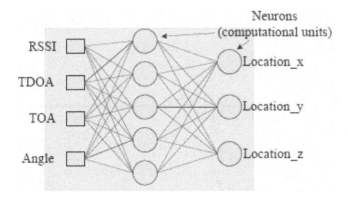

FIGURE 4.1 Node Localization Sample in WSNs.

points, making this an efficient algorithm for WSNs, coupled with the linked readings of adjacent nodes. It was shown by Beyer et al. (36) that for high-dimensional spaces, the k-NN algorithm can provide incorrect results as the samples of data have distances that seem invariant. In WSNs, the query-processing subsystem is the most critical implementation of the KNN algorithm (e.g., (22, 23)).

Decision tree (DT): This method falls under the classification category for predicting information labels by iterating entry information over a knowledge tree (37). In this procedure, the properties of the purposes are compared to the conditions for the decision toward a particular category. The literature is very rich in approaches that have used this algorithm to solve the multiple issues of WSN design (48).

For example, by identifying a variety of dangerous frameworks, like loss rate, dishonesty rate, mean time to failure, and mean time to restore, DT gives an effective idea for producing link reliability in WSNs. Nondeterministic polynomial-time complete (NP)-complete DT works with linearly separable data and the method of constructing optimal learning trees.

Neural networks (NNs): This training set is built by various interconnected chains used for decision-making units to identify nonlinear functions (9). For WSNs, it is not common to use NNs, and the reason is because of their use of huge computational resources for knowing weights of the network and high overhead management. It is possible for NNs to learn various different outputs and decision borders at the same time in centralized solutions (38) that are made.

In Figure 4.2 as an example of an NN application in sensor nodes, we consider the problem of localization of a node. The location of the node relies on the transmission angle and the distance from the anchor nodes of the received signals (39). These indicators include the received signal strength indicator (RSSI) obtained, the time of arrival (TOA) and the time gap of arrival (TDOA). After directed learning, the NN provides an average node placement as vector-valued coordinates in three-dimensional spaces. Self-organizing maps and LVQs include related NN

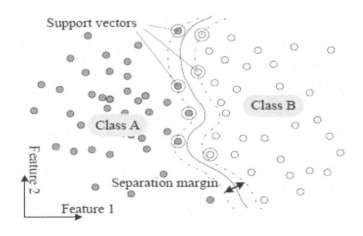

FIGURE 4.2 SVMs.

algorithms (refer to Kohonen (40); for the creation of these approaches, also refer to Kohonen (40)). Adding to this, it is among the major aspects of NNs for large data flexibility and reduced dimensions (41).

SVM, which is Support Vector Machine, is one ML algorithm that, using named class labels, learns to classify data points (42). For example, a node's malicious activity can be found out if temporal and spatial data co-relations are examined using an SVM. The SVM separates given are of space into different sections given by WSN observations as points in the function space.

These sections are segregated by margins by way of inclusion and are recognized by the side gaps from which they fall (49). An alternative solution to the non-convex and unregulated optimization of the multilayer neural circuit is known by an algorithm that helps in optimizing a quadratic function, which involves linear limitations (i.e., the limitation of building a collection of hyperplanes) (37). Protection (e.g., (31, 33, 43–45) and position are the SVM applications in WSNs (e.g., (46–48)). Please refer to Steinwart and Christmann (42) for details on SVMs.

4.3 FUNCTIONAL CHALLENGES

The node sensor's memory and processing limits, changes in topology, link failures, and decentralized governance in the WSN architecture need to be noted. To overcome multiple technical challenges in the network of wireless sensors, ML paradigms have been built, including real-time energy-saving routing, query-processing node clusters, and others (52, 53):

A. Specific opinions like fault tolerance, robustness, and the like (6) have to be considered in constructing a routing protocol for WSNs. They have reduced processing capability, less memory, and considerably less bandwidth. A routing issue is formalized as a graph G = (V,E); here, V becomes the set of all nodes, and E becomes a set of bidirectional channels of communication linking the nodes. To use this method, the regulatory control is useful for discovering minimal cost (path) for finding the correct edges of the graph. This is a T = (V,E) tree, whereby its vertices compose the root node simultaneously. Even when the entire topology is known, resolving a tree of ideal data aggregation is found NP-hard (54, 55).

Artificial intelligence helps with understanding from the history of a network's sensors the learning of efficient routing and the habituating into the dynamic environment. It is necessary to characterize the benefits as follows.

Artificial intelligence is capable of investigating the best routing paths that save energy and increase the life span of dynamically evolving WSNs. (56). It minimizes the difficulty of a standard routing issue by splitting the route into smaller problems with sub-routing. Nodes formulate the graph structures in each sub-problem by the following means:

Figures 4.3a and 4.3b use a graph and the standard dynamic routing transmission range to highlight this simple, basic routing problem. Finite routes are found

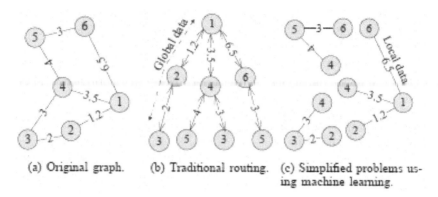

(a) Original graph.   (b) Traditional routing.   (c) Simplified problems us-
                                                    ing machine learning.

FIGURE 4.3   A Sample of a Sensor Grid Routing Problem.

by sharing route details with one another. Figure 4.3c shows how the ML algorithm minimizes the difficulty of the conventional problem by just reducing it.
The route procedures will be done separately by each node to assess the channels
to be used and the optimum transmission. In this subsection, for a nearly optimal
routing decision, as discussed earlier, such a mechanism provides very low computational complexity (57, 59). The "scalability" parameter shows the ability of
large-scale network solutions to route data (58).

A common framework for modeling sensor data was introduced by Guestrin et al.
(64). This architecture depends on the nodes matching their own measurements to
suit a global function. The nodes are used as the weighted components to perform
a kernel linear regression. Kernel functions are used here to permit the process of
manipulating data. This structure finds that the readings of many sensors are strongly
co-related, hence resulting in a minimized overhead of contact for detecting the sensor data structure. The findings obtained serve as an essential key for using linear
regression approaches to establish a distributed learning system for non-wired networks. The key advantages of using this method are the fitting outcomes and minimal overhead of the process of learning. Nonlinear and dynamic functions, however,
cannot be taught (59, 60).

Barbancho et al. (65) implemented "sensor intelligence routing" (SIR) to detect
optimal routing paths by using unsupervised learning, as shown in Figure 4.4. To
form the backbone of the network and the shortest paths in the network from a base
station to any node, SIR adds a slight change to the Dijkstra algorithm. During
route training, the second-layer neurons compete to serve more weights. The cross-
layer neurons struggle to have high weights while route training within the learning
chain. The successful neuron weights and their neighbors are adjusted to better
fit the input patterns. The trained model is a highly laboratory procedure, likely
because of the neural network generation. Therefore, it must be carried out in an
inventive main station. The execution process does not provide processing costs
and can be performed on entire network. Therefore, it takes into account the QoS
specifications during the updation process. The biggest hurdles to introducing such

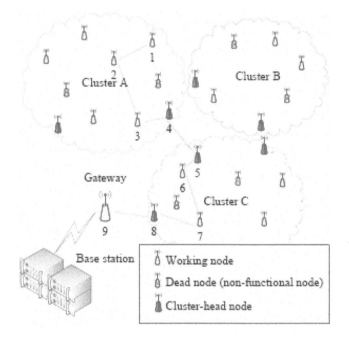

FIGURE 4.4 Information Collection Sample (clustered architecture).

an algorithm are the algorithm's complexity and the overhead of the assembly system of the event (61).

For multicast routing, one node transmits the original post to various receivers. Sun et al. (62) show that multicast routing is improved by using Q-learning algorithm in networks that are not wired. In essence, this algorithm ensures the allocation of reliable information. For different nodes, a mobile ad hoc network can have nodes that are heterogeneous in nature with separate capabilities. In comparison, retaining a universal understanding of the complete design of the network is not feasible. In two steps, the multicast paths are computed. The first stage is "Enter Query Forward," which identifies an ideal path and updates the Q-values of the Q-learning method (a prediction of potential rewards).

The second level, called "Join Reply Backward," creates the optimum route for multicast transmissions to be permitted (66). When Q-learning is applied, optimized result in flexibility and search limits of routes can be achieved (67). The 3.1- to 10.6-GHz (7500 MHz of the spectrum) frequency band has indeed been allocated for the use of ultra-wideband (UWB) (68) Federal Communications Commission. UWB is a technique that uses a large range of frequency bands with relatively low power to transmit bulky data over short distances. Dong et al. (63) used a similar concept was used to improve regional routing in sensor networks fitted with UWB (63). The protocol "Reinforcement Learning-Based Geographic Routing" reflects the vitality and delay of the sensor node in the form of metrics to find the "learning-reward function". The UWB technology uses this hierarchical geographic routing to classify the positions of the nodes. In addition, every node makes use of the "simple look-up" table to store its nearby nodes' data, whereby cluster heads are only available along with UWB devices (because the neighbour's location and energy are highly necessary). To find the efficient routing behavior, the information is shared with nodes using short "hello" messages (89). The primary advantage of using support learning in routing is that an appropriate routing solution does not require knowledge of the global network structure (90, 91).

4.3.1 CLUSTERING AND DATA AGGREGATION

For maximum vital sensor networks, transporting all data in a direct way to the sink is not an optimal solution (69). An effective method is communicating the data to a cluster head that consolidates and transmits data to sink because of all the sensors within its cluster. This reduces energy consumption. Efficient cluster head selection is addressed by several works, such as in Crosby et al. (70), Kim et al. (71), and Soro and Heinzelman (72). Abbasi and Younis (73) present the task scheduling aggregation of cluster-based-related algorithms in WSNs. Some damaged nodes are eliminated from this network in this situation. These defective nodes can cause inaccurate readings which can impair the overall network operation accuracy (92, 93). Essentially, ML algorithms boost the node classifier and data aggregation process as follows:

- Artificial intelligence in sensor nodes to lightweight data is applied locally by successfully extracting similarities and differences in different sensor readings (e.g., from faulty nodes).
- ML is implemented to accurately find cluster head, where right choice of cluster head can minimize energy consumption and improve the network's longevity (94).

Hongmei et al. (74) addressed the creation of neural network self-managed clusters. This scheme tackles the clustering issue with short transmission radii in a large-scale network in which centralized algorithms cannot operate effectively. However, the performance of this algorithm for large transmission radii is the same as for centralized algorithms.

To have a solution to the cluster head election problem, Ahmed et al. (75) used a decision tree algorithm that improves the overall efficiency of cluster head choices as opposed to low-energy adaptive clustering hierarchy (82).

A Gaussian process (GP) can be parameterized using mean and covariance functions (a mix of random variables). Ertin (76) proposes a technique, a Gaussian probability process regression, for implementing probabilistic models. Comparatively, depending on its importance, Kho et al. (77) also applied this technique for sampling data. A trade-off regarding the cost of the computation required and finding the optimal solution was studied with a focus on energy consumption (77). In general, Gaussian control systems are implemented when compared to the minimalized data sets (fewer than 1000 training data) to find smooth functions (50). WSN programmers have to take into account the high computation of methods with large-scale networks.

For a high-dimension to a low-dimension mapping of spaces, the SOM algorithm is a competitive, unsupervised learning method. Also, "cluster-based self-organizing data aggregation" (CODA) has been proposed (51). The nodes had the ability to use a self-organizing algorithm to identify the aggregated data.

The use of WSNs for the distributed detection of environmental phenomena was investigated by Krishnamachari and Iyengar (86). The reading might be regarded as not accurate if the values stay above a given threshold. This research uses Bayesian centralized learning that helps reduce 95% of the faults and knows the region of the incident. In brief, for the Bayesian dispersed algorithm proposed by Krishnamachari and Iyengar (86), these improvements lead to enhanced error and capacity estimates. Furthermore, Zappi et al. (87) developed a real-time motion classification method using WSNs that helps detect the accurate body's gesture and motion. Just use an ultrasonic sensor with positive, neutral, and null axes. All these mentioned methods are used in an hidden Markov model (HMM) to find the action of every sensor; the nodes, which spread throughout the body, initially sense organ motion. Sensing stimulation and collection are based on the possible contribution of the sensor to the classifier's accuracy. A Näive Bayes classifier is used to generate a final gesture decision that combines node predictions that are not dependent, which increases the posterior probability (96).

In fire protection and emergency systems, WSNs have been actively used. Moreover, although it costs much less, using WSN technology in the fire detection in forests will produce a more efficient solution than will satellite-based solutions. Yu et al. (24) proposed a forest fire detection system using the NN approach. Cluster heads receive the distributed data from which relevant data are applied for the decision-maker. The challenge here is that the classification role and system center are less subject to interpretation when implementing these kinds of machines.

In WSNs, KNN query has been regarded to be the optimal query-processing technique. For example, Winter et al. (22) created a solution to the problem by the "K-NN Boundary Tree" (KBT), which is the KNN algorithm. Each node will decide its KNN search region when liquid is collected. The powerful query architecture utilized by Winter et al. (22) has been correspondingly expanded by Jayaraman et al. (23). A query-processing system is "3D-KNN" for WSNs that adopts the KNN algorithm. That approach limits the search region to joining nodes in a three-dimensional space at least KNN. To maximize the KNN, signal-to-noise ratio (SNR) and diversification are needed. The disadvantages of such fork-processing algorithms are the need for memory interface footprints to store each obtained data and great delay in large-scale processing.

Bahrepouret (25) implemented detection and recognition for disaster prevention. One of the applications includes detecting fire at buildings. The result is taken by the nodes that are reputed to be on a high scale.

Malik et al. (88) streamlined the conventional enquiring mind using data attributes and PCA, therefore decreasing the optimality. PCA was helpful among all the correlated data sets to dynamically find important attributes.

4.4 NONFUNCTIONAL CHALLENGES

Partial conditions will be included in requirements that will have no application to the system's low-level operational behavior. For example, WSN designers will have to make sure that the suggested fix is always able to provide the latest information. A detailed overview of recent advances in ML to meet nonfunctional criteria such as reliability in WSNs is given in this chapter. Furthermore, this section also demonstrates some fascinating attempts to implement advanced WSN applications. Such studies may encourage researchers to explore ML techniques in a variety of WSN applications.

4.4.1 Security and Anomaly Intrusion Detection

The limited resource constraints (12) are the key obstacle to the implementation of protection techniques in WSNs. Moreover, by adding false observations to the network, some attack methods seek to generate unintended, erroneous information.

To detect outlying and misleading measurements, ML algorithms have been employed. At about the same time, by examination of well-known malicious attacks and weaknesses, several threats have been found. Basically, the following profits will result in WSN security upgrades by applying ML techniques:

- Save the power of the node and dramatically extend the lifetime of WSN by avoiding the outlier transmission, therefore resulting in unstable results
- Improve the network's reliability by removing unreliable and malicious readings. Adding to this is that it will prevent the detection of unwanted data that are, therefore, translated into substantial, sometimes crucial behavior.
- E-learning of web-attacks, vulnerabilities, and detection

Here, various ML techniques in WSNs are addressed.

Janakiram et al. (29) used Bayesian belief networks (BBNs) for finding outliers. Since most node neighbours might have equivalent readings it's also fair to make use of this phenomena to construct conditional relations between node readings. BBNs infer that any potential outliers in the data collected were detected by the conditional relations between the observations. In addition, to evaluate missing values, this method can be used.

In outlier detection using KNN, the mean price of the k-nearest nodes will also be substituted for any missing nodes' readings. However, to store all documented readings from the user's environment, a nonparametric, KNN-based algorithm includes substantial memories.

In black hole attacks, when malicious users receive "Path Query" (RREQ) messaging, the attacker node sends deceptive "Routing Reply" (RREP) messages, implying the ways to locate the destinations. Source nodes then skip the routing discovery process and ignore other RREP messages. All network messages would then be dropped by malicious nodes, whereas the nodes believe the data have been sent to the target. The method of preventing packet-dropping attacks based on a one-class SVM classifier was introduced by Kaplantzis et al. (31). The gradual shift is able to detect black hole attacks and selective forwarding attacks. Effectively, the malicious nodes of the network are analyzed using routing data, bandwidth and hop count.

With the usage of quarter-sphere centralized at the beginning, the disadvantage of huge computational needs of conventional SVM can be reduced. The identification technique of a one-class quarter-sphere SVM anomaly was developed by Rajasegarar et al. (32), for example. The objective is to recognize data anomalies, therefore reducing communication overhead. The creation of an online outlier control strategy using the SVM quarter-sphere was tackled by Yang et al. (43). By means for eliminating the complexity of available techniques by means of SVM, the method of unsupervised learning explores local data. The method applied in Rajasegarar et al. (32) is identical to this outlier detector. A computer-intelligent problem-solving framework bio-inspired immunity systems is the algorithm for artificial immunity (78). The human immunity system spontaneously creates antibodies for fighting the antigen by implementing cell fission. In short, the oblivious approach was developed (pre-processing phase) for sensor data for the SVM to detect intruders. In addition, the spatial or temporal correlations of the readings obtained were also investigated by Zhang et al. (45) to create an outlier detection mechanism using a learning algorithm for one SVM class. This paper applied an elliptic one-class SVM that in traditional SVM methods are finding solutions by implementing linear optimization rather than the quadratic optimization problem. Performance management and understanding nonlinear issues are the key advantages of these SVM-based strategies (42).

4.4.2 QoS, Data Integrity, and Fault Service

QoS discovery pledges in elevation importance transfer of real-time events and information. In addition to the distribution of queries, potential multi-hop data transfers for any user are possible within the context of WSNs (79). This research used an elliptical one-class SVM that finds solutions using linear optimization. The benefits of such SVM techniques are performance management and understanding nonlinear and complex issues. Combined with random network technology, these problems present an important challenge for such networks to design precise algorithms. The formal requirements for QoS have been looked and checked in WSNs (80).

We study the latest attempts to use machine learning techniques to meet the following specific restrictions on QoS and data integrity. This has various advantages:

- To find various kinds of streams, different ML classifiers are used, thus removing flow-aware management techniques.
- The QoS warranty, data integrity and fault detection specifications are highly dependent on network service and application. ML approaches can handle this while making effective use of resources, primarily bandwidth and power usage. Due to various factors such as signal fluctuations and interference, quality measurement instruments provide unreliable and variant readings in various environments (81).

A technique, proposed by Osborne et al. (83), which is a real-time technique, is used for evaluating nodes (in series), which has the ability to handle tasks, such as evaluating the readings of sensors, how accurate it can be and finding missing readings. Centered on the Gaussian probabilistic method, which is trained to reuse previous experience, this algorithm provides an iterative implementation.

Ouferhat and Mellouk (84) applied a QoS scheduler built on Q-learning techniques for dynamic multimedia sensor networks. This scheduler greatly increases the throughput of the network by reducing the latency of transmission. In contrast, Seah et al. (85) considered the coverage of WSNs as a QoS metric reflecting how optimally these regions of interest would be noted.

These areas will be discussed in the future. It should be noted that in the previously discussed QoS processes, energy recycling has not been considered. Alternatively, Hsu et al. (83) implemented a QoS-aware power management system, namely, "Reinforcement Learning–Based QoS-Aware Power Management" (RLPM) with capabilities of producing energy.

This device will habituate the complex energy levels of the sensor nodes. To achieve QoS knowledge and to monitor the duty cycle of nodes under energy constraint, QoS-aware RLPM uses reinforcement learning. In addition, "Multi-Agent Reinforcement Learning Based on Multi-Hop Mesh Cooperative Communication" (MRL-CC) was designed by Liang et al. (53) to be a "Multi-Agent Reinforcement Learning QoS" provisioning structure modeling method in WSNs.

Basically, to analyze the data effectively, the MRL-CC is adopted in a cooperative manner. Also, MRL-CC, helps in knowing the traffic load on any network.

4.4.3 MISCELLANEOUS APPLICATIONS

Here, numerous special research implementations are addressed.

Two researchers have proposed the distributed independent reinforcement learning (DIRL) algorithm (80), which uses available knowledge and application limitations to improve different tasks by reducing energy consumption. Each and every node finds and learns the minimal requirements needed to implement the planned tasks to optimize its potential rewards and find optimal parameters for the intended usage. Consider an object recognition program as a typical event, as shown in Figure 4.5. To optimize the network life span, these operations must be carried out at any priority, whereby the network does not have any priority information and there exists no static plan for any of the activities. For example, when starting to take samples, any sensor node has no knowledge about how/at which moment the object will be transferred. Therefore, here, a Q-learning algorithm is implemented; then a set price for a mission after which the DIRL algorithm is used to produce the necessary understanding. In many applications, WSNs have been used, such as environmental and ecosystem monitoring (81). Because most node operations must be compatible with each other, the synchronization of nodes is essential. In addition, the architecture of techniques for WSNs will have to take into account the restricted resource limitations (94).

In air-quality monitoring using neural networks, a neural method has been proposed by Postolache et al. to calculate air pollution percentage by cost-friendly gas sensor devices, thereby minimizing the influence of temperature and pressure on sensed data (92, 93). To determine the quality of air and concentration of gas, NNs

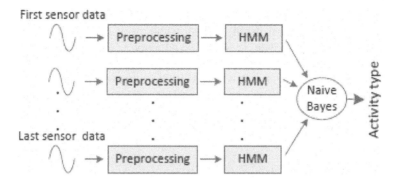

FIGURE 4.5 Human Activity Recognition Using HMM and the Näive Bayes Classifier.

are used along with JavaScript here. Consequently, web servers and end-user devices may delegate processing to the solution (90, 91).

4.5 FUTURE APPLICATIONS ON WSNS USING ML

In practice, to maintain the desired detection precision, many sensors are typically required. This poses conceptual change with so many challenges, such as network design and problems with synchronization. Because a large amount of the energy (80%) of the nodes is used when transmitting and obtaining data (83), compression of data and reduction techniques (dimensional) is used for minimizing transmission and thereby increase the life span of the network. Because of huge computational, storage needs, data compression approaches, which are old, can result in unwanted energy usage. The trade-off of energy use in transmitting data was researched by Lin et al. (78), and compression was investigated by Barr and Asanovi'c. Using 485–1267 ADD guidance (78), in here, the data compression efficiency computed is 1-bit data reduction. Corona identification routing using WSNs would also be considered for future work (97).

Distributed and adaptive ML techniques for WSNs and decentralized ML techniques fit finite devices that include WSNs. Distributed learning methods need less computing power and a smaller footprint (in the case of memory) compared to centralized learning algorithms. These allow nodes to change their current performance to rapidly.

In developing better algorithms and techniques for WSNs, energy saving is a key problem. Involving two strategies, like routing protocol design and recognizing nonfunctional habits, this design objective can be achieved. This algorithm is developed mostly using ML algorithms, as discussed in the survey. In the case of congestion, it can be detected and avoided by the transient plane (95), which has the mechanisms and possibility to employ in WSN that is based on SDN.

4.6 CONCLUSION

WSNs vary in different ways from conventional networks, requiring protocols and resources that overcome specific challenges and limitations. WSNs therefore need creative keys for routing that are sensitive to energy, real-time routing, protection, arrangement, localization, bunching of nodes, clumps of data, proof of identity of faults and the integrity of data. In short, the adoption of ML algorithms in sensor networks (wireless) must take into account the network's limited resources. As well as the range of thematic analysis is a method for learning, which helps find a solution easily. In addition, various issues remain unsolved, for example, the development of lightweight and distributed communication methods and the adoption of the wireless sensor ML resource network management problem.

REFERENCES

1. Ayodele, T. O. (2010). Introduction to Machine Learning. In *New Advances in Machine Learning*. InTech, https://www.intechopen.com/books/new-advances-in-machine-learning.
2. Duffy, A. H. (1997). The "What" and "How" of Learning in Design. *IEEE Expert*, 12(3), 71–76.
3. Langley, P., and H. A. Simon. (1995). Applications of Machine Learning Andrule Induction. *Communications of the ACM*, 38(11), 54–64.
4. Paradis, L., and Q. Han. (2007). A Survey of Fault Management in Wireless Sensor Networks. *Journal of Network and Systems Management*, 15(2), 171–190.
5. Krishnamachari, B., D. Estrin, and S. Wicker. (2002). The Impact of Data Aggregation in Wireless Sensor Networks. *22nd International Conference on Distributed Computing Systems Workshops*, 575–578.
6. Al-Karaki, J., and A. Kamal. (2004). Routing Techniques in Wireless Sensor Networks: A Survey. *IEEE Wireless Communications*, 11(6), 6–28.
7. Romer, K., and F. Mattern. (2004). The Design Space of Wireless Sensor Networks. *IEEE Wireless Communications*, 11(6), 54–61.
8. Wan, J., M. Chen, F. Xia, L. Di, and K. Zhou. (2013). From Machine-to Machine Communications Towards Cyber-Physical Systems. *Computer Science and Information Systems*, 10, 1105–1128.
9. Bengio, Y. (2009). Learning Deep Architectures for AI. *Foundations and Trends in Machine Learning*, 2(1), 1–127.
10. Di, M., and E. M. Joo. (2007). A Survey of Machine Learning in Wireless Sensor Networks from Networking and Application Perspectives. In *6th International Conference on Information, Communications Signal Processing*, 1–5.
11. Förster, A., and M. Amy. (2011). Machine Learning Across the WSN Layers. *InTech Open*, doi:10.5772/10516.
12. Zhang, Y., N. Meratnia, and P. Havinga. (2010). Outlier Detection Techniques for Wireless Sensor Networks: A Survey. *IEEE Communications Surveys& Tutorials*, 12(2), 159–170.
13. Hodge, V. J., and J. Austin. (2004). A Survey of Outlier Detection Methodologies. *Artificial Intelligence Review*, 22(2), 85–126.
14. R. Kulkarni, A. Förster, and G. Venayagamoorthy. (2011). Computational Intelligence in Wireless Sensor Networks: A Survey. *IEEE Communications Surveys & Tutorials*, 13(1), 68–96.
15. Das, S., A. Abraham, and B. K. Panigrahi. (2010). *Computational Intelligence: Foundations, Perspectives, and Recent Trends*. Chichester: John Wiley & Sons, Inc., 1–37.
16. Abu-Mostafa, Y. S., M. Magdon-Ismail, and H. T. Lin. (2012). *Learning from Data*. Chicago: AML Book.

17. Chapelle, O., B. Schlkopf, and A. Zien. (2006). *Semi-Supervised Learning.* Cambridge: MIT Press, 2.

18. Kulkarni, S., G. Lugosi, and S. Venkatesh. (1998). Learning Pattern Classification-a Survey. *IEEE Transactions on Information Theory,* 44(6), 2178–2206.

19. Morelande, M., B. Moran, and M. Brazil. (2008). Bayesian Node Localisation in Wireless Sensor Networks. *IEEE International Conference on Acoustics, Speech and Signal Processing,* 2545–2548.

20. Lu, C. H., and L. C. Fu. (2009). Robust Location-Aware Activity Recognition Using Wireless Sensor Network in an Attentive Home. *IEEE Transactions on Automation Science and Engineering,* 6(4), 598–609.

21. Shareef, A., Y. Zhu, and M. Musavi. (2008). Localization Using Neural Networks in Wireless Sensor Networks. *Proceedings of the 1st International Conference on Mobile Wireless Middleware, Operating Systems, and Applications,* 1–7.

22. Winter, J., Y. Xu, and W. C. Lee. (2005). Energy Efficient Processing of k Nearest Neighbor Queries in Location-Aware Sensor Networks. *2nd International Conference on Mobile and Ubiquitous Systems: Networking and Services,* 281–292.

23. Jayaraman, P. P., A. Zaslavsky, and J. Delsing. (2010). Intelligent Processing of k-Nearest Neighbors Queries Using Mobile Data Collectors in a Location Aware 3D Wireless Sensor Network. *Trends in Applied Intelligent Systems,* 260–270.

24. Yu, L., N. Wang, and X. Meng. (2005). Real-Time Forest Fire Detection with Wireless Sensor Networks. *International Conference on Wireless Communications, Networking and Mobile Computing,* 2, 1214–1217.

25. Bahrepour, M., N. Meratnia, M. Poel, Z. Taghikhaki, and P. J. Havinga. (2010). Distributed Event Detection in Wireless Sensor Networks for Disaster Management. *2nd International Conference on Intelligent Networking and Collaborative Systems. IEEE,* 507–512.

26. Kim, M., and M. G. Park. (2009). Bayesian Statistical Modeling of System Energy Saving Effectiveness for MAC Protocols of Wireless Sensor Networks. In *Software Engineering, Artificial Intelligence, Networking and Parallel/Distributed Computing,* ser. Studies in Computational Intelligence. Berlin, Heidelberg: Springer, vol. 209, 233–245.

27. Shen, Y. J., and M. S. Wang. (2008). Broadcast Scheduling in Wireless Sensor Networks Using Fuzzy Hopfield Neural Network. *Expert Systems with Applications,* 34(2), 900–907.

28. Kulkarni, R. V., and G. K. Venayagamoorthy. (2009). Neural Network Based Secure Media Access Control Protocol for Wireless Sensor Networks. In *Proceedings of the 2009 International Joint Conference on Neural Networks,* ser. IJCNN'09. Piscataway, NJ: IEEE Press, 3437–3444.

29. Janakiram, D., V. Adi Mallikarjuna Reddy, and A. Phani Kumar. (2006). Outlier Detection in Wireless Sensor Networks Using Bayesian Belief Networks. *1st International Conference on Communication System Software and Middleware. IEEE,* 1–6.

30. Branch, J. W., C. Giannella, B. Szymanski, R. Wolff, and H. Kargupta. (2013). In-Network Outlier Detection in Wireless Sensor Networks. *Knowledge and Information Systems,* 34(1), 23–54.

31. Kaplantzis, S., A. Shilton, N. Mani, and Y. Sekercioglu. (2007). Detecting Selective Forwarding Attacks in Wireless Sensor Networks Using Support Vector Machines. *3rd International Conference on Intelligent Sensors, Sensor Networks and Information. IEEE,* 335–340.

32. Rajasegarar, S., C. Leckie, M. Palaniswami, and J. Bezdek. (2007). Quarter Sphere Based Distributed Anomaly Detection in Wireless Sensor Networks. *International Conference on Communications,* 3864–3869.

33. Snow, A., P. Rastogi, and G. Weckman. (2005). Assessing Dependability of Wireless Networks Using Neural Networks. *Military Communications Conference. IEEE*, 5, 2809–2815.

34. Moustapha, A., and R. Selmic. (2008). Wireless Sensor Network Modeling Using Modified Recurrent Neural Networks: Application to Fault Detection. *IEEE Transactions on Instrumentation and Measurement, 57(5), 981–988.*

35. Wang, Y., M. Martonosi, and L. S. Peh. (2007). Predicting Link Quality Using Supervised Learning in Wireless Sensor Networks. *ACM Sigmobile Mobile Computing and Communications Review*, 11(3), 71–83.

36. Beyer, K., J. Goldstein, R. Ramakrishnan, and U. Shaft. (1999). When Is "Nearest Neighbor" Meaningful? In *Database Theory*. Cham: Springer, 217–235.

37. Ayodele, T. O. (2010). Types of Machine Learning Algorithms. In *New Advances in Machine Learning*. InTech, https://www.intechopen.com/books/new-advances-in-machine-learning/types-of-machine-learning-algorithms.

38. Lippmann, R. (1987). An Introduction to Computing with Neural Nets. *ASSP Magazine, IEEE*, 4(2), 4–22.

39. Dargie, W., and C. Poellabauer. (2010). *Localization*. New York: John Wiley & Sons, Ltd., 249–266.

40. Kohonen, T. (2001). *Self-Organizing Maps*, ser. Springer Series in Information Sciences. Berlin, Heidelberg: Springer, vol. 30.

41. Hinton, G. E., and R. R. Salakhutdinov. (2006). Reducing the Dimensionality of Data with Neural Networks. *Science*, 313(5786), 504–507.

42. Steinwart, I., and A. Christmann. (2008). *Support Vector Machines*. New York: Springer.

43. Yang, Z., N. Meratnia, and P. Havinga. (2008). An Online Outlier Detection Technique for Wireless Sensor Networks Using Unsupervised Quarter-Sphere Support Vector Machine. *International Conference on Intelligent Sensors, Sensor Networks and Information Processing. IEEE*, 151–156.

44. Chen, Y., Y. Qin, Y. Xiang, J. Zhong, and X. Jiao. (2011). Intrusion Detection System Based on Immune Algorithm and Support Vector Machine in Wireless Sensor Network. In *Information and Automation*, ser. Communications in Computer and Information Science. Berlin, Heidelberg: Springer, vol. 86, 372–376.

45. Zhang, Y., N. Meratnia, and P. J. Havinga. (2013). Distributed Online Outlier Detection in Wireless Sensor Networks Using Ellipsoidal Support Vector Machine. *Ad Hoc Networks*, 11(3), 1062–1074.

46. Kim, W., J. Park, and H. Kim. (2010). Target Localization Using Ensemble Support Vector Regression in Wireless Sensor Networks. *Wireless Communications and Networking Conference*, 1–5.

47. Tran, D., and T. Nguyen. (2008). Localization in Wireless Sensor Networks Based on Support Vector Machines. *IEEE Transactions on Parallel and Distributed Systems*, 19(7), 981–994.

48. Yang, B., J. Yang, J. Xu, and D. Yang. (2007). Area Localization Algorithm for Mobile Nodes in Wireless Sensor Networks Based on Support Vector Machines. In *Mobile Ad-Hoc and Sensor Networks*. Cham: Springer, 561–571.

49. Box, G. E., and G. C. Tiao. (2011). *Bayesian Inference in Statistical Analysis*. New York: John Wiley & Sons, vol. 40.

50. Rasmussen, C. E. (2006). Gaussian Processes for machine Learning. In *Adaptive Computation and Machine Learning*. Cambridge: MIT Press, Citeseer.

51. Lee, S., and T. Chung. (2005). Data Aggregation for Wireless Sensor Networks Using Self-Organizing Map. In *Artificial Intelligence and Simulation*, ser. Lecture Notes in Computer Science. Berlin, Heidelberg: Springer, vol. 3397, 508–517.

52. Masiero, R., G. Quer, D. Munaretto, M. Rossi, J. Widmer, and M. Zorzi. (2009). Data Acquisition Through Joint Compressive Sensing and Principal Component Analysis. *Global Telecommunications Conference. IEEE*, 1–6.

53. Masiero, R., G. Quer, M. Rossi, and M. Zorzi. (2009). A Bayesian Analysis of Compressive Sensing Data Recovery in Wireless Sensor Networks. *International Conference on Ultra Modern Telecommunications Workshops*, 1–6.

54. Rooshenas, A., H. Rabiee, A. Movaghar, and M. Naderi. (2010). Reducing the Data Transmission in Wireless Sensor Networks Using the Principal Component Analysis. *6th International Conference on Intelligent Sensors, Sensor Networks and Information Processing. IEEE*, 133–138.

55. Macua, S., P. Belanovic, and S. Zazo. (2010). Consensus-Based Distributed Principal Component Analysis in Wireless Sensor Networks. *11th International Workshop on Signal Processing Advances in Wireless Communications*, 1–5.

56. Tseng, Y. C., Y. C. Wang, K. Y. Cheng, and Y. Y. Hsieh. (2007). iMouse: An Integrated Mobile Surveillance and Wireless Sensor System. *Computer*, 40(6), 60–66.

57. Li, D., K. Wong, Y. H. Hu, and A. Sayeed. (2002). Detection, Classification, and Tracking of Targets. *IEEE Signal Processing Magazine*, 19(2), 17–29.

58. Kanungo, T., D. M. Mount, N. S. Netanyahu, C. D. Piatko, R. Silverman, and A. Y. Wu. (2002). An Efficient k-Means Clustering Algorithm: Analysis and Implementation. *IEEE Transactions on Pattern Analysis and Machine Intelligence*, 24(7), 881–892.

59. Jolliffe, I. T. (2002). *Principal Component Analysis*. New York: Springer.

60. Feldman, D., M. Schmidt, C. Sohler, D. Feldman, M. Schmidt, and C. Sohler. (2013). Turning Big Data into Tiny Data: Constant-Size Core Sets for k-Means, PCA and Projective Clustering. *SODA: Symposium on Discrete Algorithms*, 1434–1453.

61. Watkins, C., and P. Dayan. (1992). Q-Learning. *Machine Learning*, 8(3–4), 279–292.

62. Sun, R., S. Tatsumi, and G. Zhao. (2002). Q-MAP: A Novel Multicast Routing Method in Wireless Ad hoc Networks with Multiagent Reinforcement Learning. *Region 10 Conference on Computers, Communications, Control and Power Engineering*, 1, 667–670.

63. Dong, S., P. Agrawal, and K. Sivalingam. (2007). Reinforcement Learning Based Geographic Routing Protocol for UWB Wireless Sensor Network. *Global Telecommunications Conference. IEEE*, 652–656.

64. Guestrin, C., P. Bodik, R. Thibaux, M. Paskin, and S. Madden. (2004). Distributed Regression: An Efficient Framework for Modeling Sensor Network Data. *3rd International Symposium on Information Processing in Sensor Networks*, 1–10.

65. Barbancho, J., C. León, F. Molina, and A. Barbancho. (2007). A New QoS Routing Algorithm Based on Self-Organizing Maps for Wireless Sensor Networks. *Telecommunication Systems*, 36, 73–83.

66. Scholkopf, B., and A. J. Smola. (2001). *Learning with Kernels: Support Vector Machines, Regularization, Optimization, and Beyond*. Cambridge: MIT Press.

67. Kivinen, J., A. Smola, and R. Williamson. (2004). Online Learning with Kernels. *IEEE Transactions on Signal Processing*, 52(8), 2165–2176.

68. Aiello, G., and G. Rogerson. (2003). Ultra-Wideband Wireless Systems. *IEEE Microwave Magazine*, 4(2), 36–47.

69. Rajagopalan, R., and P. Varshney. (2006). Data-Aggregation Techniques in Sensor Networks: A Survey. *IEEE Communications Surveys & Tutorials*, 8(4), 48–63.

70. Crosby, G., N. Pissinou, and J. Gadze. (2006). A Framework for Trust-Based Cluster Head Election in Wireless Sensor Networks. *2nd IEEE Workshop on Dependability and Security in Sensor Networks and Systems*, 10–22.

71. Kim, J. M., S. H. Park, Y. J. Han, and T. M. Chung. (2008). CHEF: Cluster Head Election Mechanism Using Fuzzy Logic in Wireless Sensor Networks. *10th International Conference on Advanced Communication Technology*, 1, 654–659.

72. Soro, S., and W. Heinzelman. (2005). Prolonging the Lifetime of Wireless Sensor Networks via Unequal Clustering. *19th IEEE International Parallel and Distributed Processing Symposium*, 4–8.

73. Abbasi, A. A., and M. Younis. (2007). A Survey on Clustering Algorithms for Wireless Sensor Networks. *Computer Communications*, 30(14), 2826–2841.

74. He, H., Z. Zhu, and E. Makinen. (2009). A Neural Network Model to Minimize the Connected Dominating Set for Self-Configuration of Wireless Sensor Networks. *IEEE Transactions on Neural Networks*, 20(6), 973–982.

75. Ahmed, G., N. M. Khan, Z. Khalid, and R. Ramer. (2008). Cluster Head Selection Using Decision Trees for Wireless Sensor Networks. *International Conference on Intelligent Sensors, Sensor Networks and Information Processing. IEEE*, 173–178.

76. Ertin, E. (2007). Gaussian Process Models for Censored Sensor Readings. *14th Workshop on Statistical Signal Processing. IEEE*, 665–669.

77. Kho, J., A. Rogers, and N. R. Jennings. (2009). Decentralized Control of Adaptive Sampling in Wireless Sensor Networks. *ACM Transactions on Sensor Networks (TOSN)*, 5(3), 19:1–19:35.

78. Lin, S., V. Kalogeraki, D. Gunopulos, and S. Lonardi. (2006). Online Information Compression in Sensor Networks. *IEEE International Conference on Communications*, 7, 3371–3376.

79. Fenxiong, C., L. Mingming, W. Dianhong, and T. Bo. (2013). Data Compression Through Principal Component Analysis Over Wireless Sensor Networks. *Journal of Computational Information Systems*, 9(5), 1809–1816.

80. Förster, A., and A. Murphy. 2009. CLIQUE: Role-Free Clustering with q Learning for Wireless Sensor Networks. *29th IEEE International Conference on Distributed Computing Systems*, 441–449.

81. Mihaylov, M., K. Tuyls, and A. Nowe. (2010). Decentralized Learning in Wireless Sensor Networks. In *Adaptive and Learning Agents*, ser. Lecture Notes in Computer Science. Berlin, Heidelberg: Springer, vol. 5924, 60–73.

82. Heinzelman, W. B. (2000). *Application-Specific Protocol Architectures for Wireless Networks*. Ph.D. dissertation, MIT Press, Cambridge.

83. Duarte, M., and Y. Eldar. (2011). Structured Compressed Sensing: From Theory to Applications. *IEEE Transactions on Signal Processing*, 59(9), 4053–4085.

84. Dempster, A. P., N. M. Laird, and D. B. Rubin. (1977). Maximum Likelihood from Incomplete Data via the EM Algorithm. *Journal of the Royal Statistical Society: Series B (Methodological)*, 1–38.

85. DeGroot, M. H. (1974). Reaching a Consensus. *Journal of the American Statistical Association*, 69(345), 118–121.

86. Krishnamachari, B., and S. Iyengar. (2004). Distributed Bayesian Algorithms for Fault-Tolerant Event Region Detection in Wireless Sensor Networks. *IEEE Transactions on Computers*, 53(3), 241–250.

87. Zappi, P., C. Lombriser, T. Stiefmeier, E. Farella, D. Roggen, L. Benini, and G. Tröster. (2008). Activity Recognition from On-Body Sensors: Accuracy Power Trade-Off by Dynamic Sensor Selection. In *Wireless Sensor Networks*. New York: Springer, 17–33.

88. Malik, H., A. Malik, and C. Roy. (2011). A Methodology to Optimize Query in Wireless Sensor Networks Using Historical Data. *Journal of Ambient Intelligence and Humanized Computing*, 2, 227–238.

89. Chen, Q., K. Y. Lam, and P. Fan. (2005). Comments on "Distributed Bayesian Algorithms for Fault-Tolerant Event Region Detection in Wireless Sensor Networks. *IEEE Transactions on Computers*, 54(9), 1182–1183.

90. Sha, K., W. Shi, and O. Watkins. (2006). Using Wireless Sensor Networks for Fire Rescue Applications: Requirements and Challenges. *IEEE International Conference on Electro/information Technology*, 239–244.

91. Liu, H., H. Darabi, P. Banerjee, and J. Liu. (2007). Survey of Wireless Indoor Positioning Techniques and Systems. *IEEE Transactions on Systems, Man, and Cybernetics, Part C: Applications and Reviews*, 37(6), 1067–1080.
92. Wang, J., R. Ghosh, and S. Das. (2010). A Survey on Sensor Localization. *Journal of Control Theory and Applications*, 8(1), 2–11.
93. Nasipuri, A., and K. Li. (2002). A Directionality Based Location Discovery Scheme for Wireless Sensor Networks. *Proceedings of the 1st ACM International Workshop on Wireless Sensor Networks and Applications. ACM*, 105–111.
94. Yun, S., J. Lee, W. Chung, E. Kim, and S. Kim. (2009). A Soft Computing Approach to Localization in Wireless Sensor Networks. *Expert Systems with Applications*, 36(4), 7552–7561.
95. Vanamoorthy, Muthumanikandan, and Valliyammai Chinnaiah. (2020). Corrigendum to: Congestion Free Transient Plane (CFTP) Using Bandwidth Sharing During Link Failures in SDN. *The Computer Journal*, bxaa024, doi:10.1093/comjnl/bxaa024.
96. Chagas, S., J. Martins, and L. de Oliveira. (2012). An Approach to Localization Scheme of Wireless Sensor Networks Based on Artificial Neural Networks and Genetic Algorithms. *10th International Conference on New Circuits and Systems. IEEE*, 137–140.
97. Thyagarajan, J., and S. Kulanthaivelu. (2020). A Joint Hybrid Corona Based Opportunistic Routing Design with Quasi Mobile Sink for IoT Based Wireless Sensor Network. *Journal of Ambient Intelligence and Humanized Computing*. doi:10.1007/s12652-020-02116-6.

5 Machine Learning Approaches in Big Data Analytics Optimization for Wireless Sensor Networks

G. Sabarmathi and R. Chinnaiyan

CONTENTS

The communication happens through numerous network modes for satisfying the needs of an individual and forms an enlarged wireless network. Simultaneously, an innumerable flow of mass data is produced through distinctive gadgets for instance sensors, laptops, mobile applications, smart applications etc. that in turn gave rise to new age called Big Data. In short it can be said that data created from enlarged wireless networks again and again are highlighted by means of varied range, massive size, real-time speed, and high significance.

G. Sabarmathi & Dr. R. Chinnaiyan

DOI: 10.1201/9781003107477-5

5.1 INTRODUCTION

The network operator has to face a challenge in operating the network efficiently with flexibility. This arises from the expanded services that demand the satisfaction of a network operator in providing the services with limited resources; it is made more complicated by manually configuring the network design and optimization. The essential key factor here is the end of network optimization, which is not an option. To reduce the operational costs, operators face concerns about the automation of networks. It is always difficult to recognize and manage a complex system like wireless networks. So we have to change these challenges to benefits by applying the correct technologies (3). Based on key performance indicators (KPIs) from multiple locations, decisions are made by the network operator by using an analysis tool. The optimization and the monitoring of networks have to access a huge volume of data generated by multiple sources through recorded information that leads to a reduction in capacity. By using the analytics on big data, it exposes the hidden patterns and the relationships among the data (4). This leads to the procurement of knowledge and unidentified data and methods for improving the performance of wireless networks from numerous points. The examples of sources of wireless big data traffic is shown in Figure 5.1.

5.1.1 BIG DATA ANALYTICS IN WIRELESS SENSOR NETWORKS

The rapid development of enlarged wireless networks has led to many challenges in designing the scalability of networks, along with their significance in varied fields, such as security, optimization, dealing with the traffic of signals, related operations, and managerial study. This entails the need for big data analytics for enlarged network services. Big data analytics involves numerous stages before the extraction of

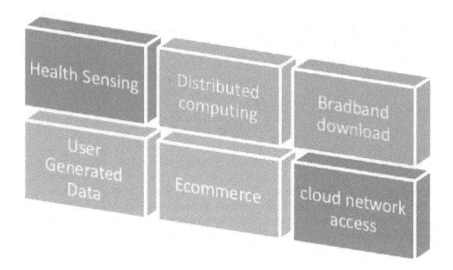

FIGURE 5.1   Examples of Sources of Wireless Big Data Traffic.

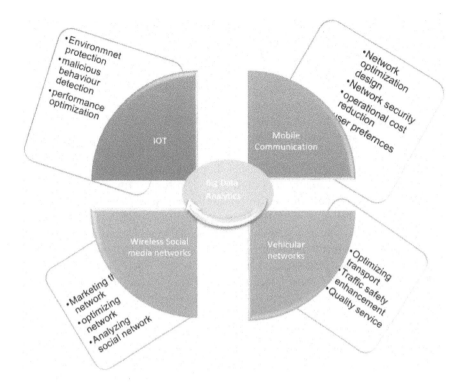

FIGURE 5.2   Need for Big Data Analytics.

hidden values, such as collection of data, filtering, storing, and finally analyzing those data.

The latest improvement in big data analytics has given rise to network optimization in wireless networks, and its necessity in today's technological world is shown in Figure 5.2. In the figure are the four main categories of data generation for wireless networks and the need for analyzing them. Let us discuss each in detail:

1. **Internet of Things (IoT):** Big data analytics in the IoT has paved the way for various usages like (a) environment protection by making correct decisions as soon as there is any vulnerability in an environment and its protection policies with the use of various sensors mounted by IoT technology (5, 30–38). It also supports instant response during emergencies in an industrial environment based on real-time analysis. (b) Malicious behavior detection can be done using the IoT; smart gadgets detect the theft in electricity and the food supply chain and ensure food safety (6). (c) Optimizing system performance with the analysis of IoT data leads to efficiency in the system's performance by predicting the accuracy in logistic data that, in turn, reduces the turnaround time (7, 8).

2. **Mobile communication:** The challenge faced in mobile communication by the network operator is increased demand by the user on the cost of service

and energy consumption. It is important that to optimize the cost and energy consumption by using big data analytics as follows: (a) User preferences in wireless communication can be improved with uninterrupted network connectivity and lowered service cost that expands the quality of service by varying optimization schemes by analyzing the data generated by both wireless services and the end user. (b) Operational cost reduction can be obtained through big data analytics, which extracts the hidden pattern and information from the wide-ranging sources of data, either instantaneous or stowed; this helps frame the strategies efficiently for making decisions. (c) For network optimization design, big data analytics benefits network optimization by mutually considering the data from software-defined networks and self-organizing networks as well as efficiently achieving network traffic (2). (d) Network security can be obtained as a benefit of big data analytics by identifying many abnormalities and malicious actions in a network.

3. **Wireless social media network:** With the era of wireless technologies and the spread of diverse wireless gadgets, enormous amounts of user data are generated from many wireless social nodes (39–42), such as media, blogs, and Twitter, through which users are connected to a wireless network virtually. Big data analytics helps wireless social media by acquiring the hidden information of the user preferences and their behavior and identifying their socioeconomic relations with the social gadgets, which helps in the following ways: (a) Marketing or promoting the network's activities to end users by way of advertisements and product promotional activities done through the reference structure's optimizing network helps with location-based services (9). (c) Analyzing social networks can be achieved through big data analytics by predicting the unusual behavior in a social mode and socioeconomic changes, if any.

4. **Vehicular Network:** The various uses of big data analytics in vehicular networks are (a) optimizing the transport structure whereby it predicts the flow of traffic and plans accordingly the lane planning and lighting in lane infrastructure. (b) Traffic safety enhancement helps the user predict the lane traffic and any changes in travel route due to any work. It also ensures safe traveling by giving live traffic information on traffic jams and assists users in altering their route. (c) Using big data analytics, quality service can be assisted in driving by giving the live information on traffic and helping users plan their route with the given information while simultaneously ensuring the safety and quality of services.

5.1.2 Types of Analytics

The progression of big data analytics has provided a way for descriptive analyses through diagnostics to predict and shine a light in the direction of prescriptive analysis. Figure 5.3a describes the varied analysis. From the diagram, it is can be seen that the wireless network presently in descriptive analysis where it visualizes what happening in the lane profile and their performance. Diagnostic analysis gives a reason for irregularities in the network functions by using the methods like correlations, deep learning, information retrieval, and others.

To predict the future, forecasting in lane patterns, congestion in the network based on live data, is done through predictive analysis. To obtain predictive results from analyzing data using varied statistical tools with machine learning (ML) methods, the data are mined to get the predicted results. Prescriptive analysis helps in the decision-making process by visualizing and slicing techniques.

This helps the network operator by analyzing various internal and external data sources and giving feedback and corresponding actions to be taken to obtain the efficiency of the data generated. As shown in Figure 5.3b, the information is categorized into internal and external based on the behavior of the data, where the data generated by the network operators and network is related to subscribers', the network's, and

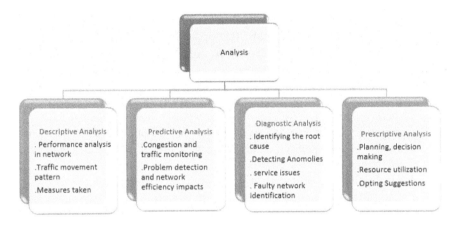

FIGURE 5.3a Various Big Data Analyses Related to Wireless Networks

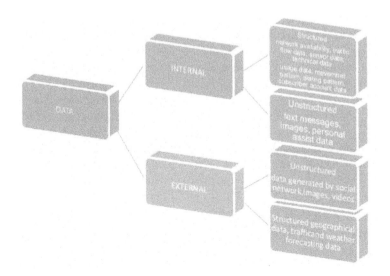

FIGURE 5.3b Big Data Analytics Data Sets and Their Sources.

third parties' data. These data are further classified as structured and unstructured data in which structured data have some relationship among the data and unstructured data do not show a relationship among the data (11, 12).

5.1.3 Drivers of Analytics Implementation

Demands on the complexity of the network traffic patterns have made the analysis of big data much needed by operators. Initially, these operators were cautious about using analytics, as numerous drivers made the operators' attitudes be stable in adopting for future usage. As a result, there arose a reliable and constant commitment to understanding the idea of network usage in procuring network optimization by means of analysis. Many drivers are used in obtaining big data analytics:

1. **COST AND SERVICE DRIVERS:** Basically, subscribers are expecting more but have less of an interest in paying. In such cases, it is necessary that network resources be optimized in an efficient way. As we know, currently, we are going toward customer-friendly services, which, in turn, make operators work towards achieving a better quality of experience (QoE) and network relationships. To maintain customers' loyalty, operators have to (a) manage network traffic based on the services offered, (b) maintain a profit along with efficiency, (c) performance and QoE improvement without raising costs, and (d) low churn.
2. **USAGE DRIVERS:** With varying network traffic patterns, customer preferences, and their usage, the customer-centric model supports the analytical services of the operators to maintain normal network traffic, with mobile devices and customer variation–based services as primary concerns. Added to this is that operators are trying their best to maintain cost-efficiency in spite of huge traffic loads. Proper use of resources is needed to maintain the stability of network services based on the load capacity.
3. **TECHNOLOGY DRIVERS:** Future network technology, such as virtualizing the network resources, slicing, computing the network edge, and the like, is ready to accept the new range of use cases along with varied wireless interfaces and layers. Network operators are calling for a booming analytics framework to coordinate virtualized network assets methodically. Analytics, again, aid network operators in appraising rationalized and broken functionality. Data analytics oscillations leave network operators to riddle out the most adequate way to slice the network and traffic, that is, the number of slices, splitting traffic across slices, and so on, including trust in the type of traffic and how it fluctuates over time and space.

5.2 ML AND ARTIFICIAL INTELLIGENCE IN WIRELESS SENSOR NETWORKS

ML and artificial intelligence (AI) are the two emerging solutions for big data, particularly in providing the solutions based on predictions in the data sets. Actually, both are interrelated such that AI devices can be used for ML-based models to make smart decisions. Application-oriented models in AI help next-generation wireless communication to a great extent in analyzing, monitoring, and enhancing. In predictive analysis, ML plays a vital role (13, 30–38), and AI suggests solutions/ideas based on the benefit of implementing the recommendations.

On the technological front, we have noticed that there are more challenges in wireless communication as there has been rapid growth in size and complexity arising from advancements in the area of networks. ML and AI are the most suitable techniques we can adopt for extracting the hidden patterns and information from the raw data. Because we know there are no historical data to depend on, the occurrences of network behavior, henceforth depending on real time data, and the predictions and suggestions are made by ML and AI with human intelligence. These two will work together with data generated from various sources and types and will predict the needed information. They will also disclose the hidden structure of the data and their relationships, which has not been predicted before by deeply studying the raw data and its insights. Based on the available tools, it is difficult for network operators to understand the hidden complexities in a wireless network; ML and AI are the future, and the algorithms related to them helps us overcome problems and gives suggestions for our future wireless network tasks. They also provide various properties in wireless networks, such as identifying the different relationships and abnormalities not normally recognized and recommending an innovative path for optimizing the network and its operations.

5.3 OPTIMIZATION TECHNIQUES

The traditional network design structure was unable to capture all the data the created problems in quality service. Mobile operators were looking for solutions that gave good-quality services to subscribers at their location by analyzing all the data captured in the network. This will reduce their time and cost and efficiently improve performance. Big data analytics was a trend-setter and solution for these operators to analyze the massive volume of data. The data analysis was able to predict the hidden pattern based on the source from networks and subscriber data that helped them optimize their network and the design structure. Figure 5.4 shows the various optimization sources in wireless networks. It is elaborated on in the following:

- User Data (UD): In this perspective, UD have the details about the device data, agreement related to service, users' affordability toward data usages (price paid per unit data rate), service quality, and the like. In all these, the UD plays a major role in optimizing and controlling the processes mentioned earlier. It is seen that whenever UD have been allotted, factors such as use of a resource, traffic control, and lane control are considered. The behavior data in UD give details on users' behaviors in using services like the duration of both audio and video calls. With the help of analytics it, easy to predict these factors and their relationships.
- User Point of View (User-P): User-P is considered a measure in wireless networks by network operators for pricing, quality related to service, and efficiency by delivering user-centric data for analysis (14). User-P is normally

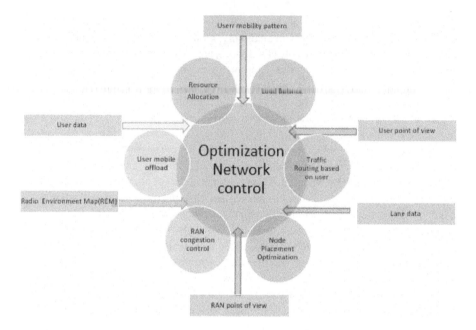

FIGURE 5.4   Optimization Data Source in Wireless Networks.

said to be the cost related to the quality of services based on many measures offered by network operators.

- Random Access Network (RAN) Point of View (RAN-P): RAN-P is the performance of RAN based on user perspective (15). Network operators analyze users' views on measures like signal intensity, network availability, and erroneous code. The devices of users' perspectives act as a network point for measuring RAN-P.
- Radio Environment Map (REM): REM plays a vital role in the congestion of a network cycle, whereby it gets the data from spectrum-sensing radio and associated information from network operators everywhere. It helps in planning, adopting new techniques/methods, and making decisions based on users' perceptions and their expectations.
- Traffic Profile (TP): TP is a profile of the traffic in a network that is based on the connection data collected over a period. We can use these connection data from our devices by exporting the data from any network-enabled device. Once we have created the profile, it helps us detect anomalies in the network traffic and suggests new traffic without any abnormalities in a network; the data collected over a period in TP is said to be a profiling time window (PTW).
- User Mobility Pattern (UMP): UMP plays a key role in designing the protocol and performance analysis in a wireless network. UMP helps us identify various users based on their arrival time, usage time, particular area of usages on a wireless network, and the quality of services used. It also helps

with predicting the future movement of the user based on the geographical locations (16).

- It is really challenging to do a study on well-balanced load system in wireless communication with many wireless users using multiple applications and services at a time. Wireless communication of the future should consider analyzing systems, users, and their services based on optimization and control over the network by the following factors (17).
- Allocating resources is important for improving the range, frequency, and power consumption in wireless network. Advanced techniques like big data analytics have been used to predict the usages of optimizing the available resource allocation. Based on user profiles, operators can predict users' needs and preferences when they are requesting the services; this helps operators optimize resources for users based on the service demand.
- Load Balance: Based on the user profiles, lanes in each node are varied, and the corresponding pattern changes based on time. Suppose the user is not happy with one node and moves to another node, causing a lane disturbance in the network; this leads to lane traffic overload and disturbs services. Manual usage of load balance by operator is not that efficient and accurate. By using big data analytics, finding the correlations among the data and predicting the load in the network and variation in traffic are easy. Accordingly, operators accomplish load balance by planning the segment capacity.
- RAN Congestion Control: In RAN congestion control, the main measures to be considered are controlling or getting rid of congestion in wireless networks. Recently, it has given an insight into innovation in this area. The foremost reason is wireless broadband deployment services and the related use of wireless gadgets. The effective use of RAN congestion helps with cost-saving in investment and offering a good QoE to users.
- Traffic routing based on users is one of the basic requirements expected from the User-P. Factors such as delays in service, interrupted network coverage, and varying ranges affect users' quality of service very badly. Based on their profiles, applications, and preferences, the solutions needed by users can be provided based on the QoE of the wireless network. The system can act proactively based on users' preferences using network optimization.
- User Mobile Offload: Owing to the maximum use of mobile data, network operators are trying to optimize network lanes by providing uninterrupted ranges and properly assuring quality of service to users. Based on traffic, operators optimize networks through Wi-Fi networks by maximizing the throughput for single user. This is done based on the network congestion and user profiles; the mobile offload is transferred to a Wi-Fi-based network by using practical approaches (18).
- Node placement optimization is needed; as we have seen, manual node planning consumes more time and is inefficient; automation for optimizing the range, data volume, and the network quality in a shorter time is needed (19). Rapidly placing small nodes in good locations is tedious work as small nodes are more in number. The older way of managing node optimization will lead to the improper placement of nodes in the network. Based on

data analytics, node optimization is efficiently monitored, with the node placement based on the user and network data. This helps with minimizing congestion and improving traffic offloading and network capacity. The geo-locator helps operators predict traffic signals in real time based on sensors.

5.3.1 ANALYTICAL TOOLS USED IN BIG DATA

For tracking down wireless traffic features, we can design a special learning unit by installing various base stations and a central unit for processing the data and analyzing them. Data analytical algorithms are inserted as key supporting factors. Various algorithms related to wireless communication and their usages are précised in Table 5.1. The table gives an overall idea of the different areas and types of algorithms, and their applications are summarized as follows:

1. **Probabilistic model:** This model uses various statistical analyses for capturing the features and changing aspects of data traffic flow. The regularly used models are time-series algorithm, linear and nonlinear random dynamic models, Markov and Kalman filter models, and geometric models. This type of model helps with predicting mobility, device association, and resource sharing.
2. **Data mining:** This area focuses on extracting information and retrieving the hidden patterns associated with the wireless data. Algorithms such as clustering, dimension reduction pattern identification, and the compression of text are used in wireless data analysis. Data mining identifies the hidden findings in the data movement from the movement history. Algorithms,

TABLE 5.1
Big Data Analytical Algorithms and Applications Based on the Major Areas.

Major Area	Analytical Algorithms	Examples
Probabilistic model	Time-series algorithm, Markov and Kalman models, geometric models	Predicting the mobility, device association, resource sharing
Data mining	Clustering, dimension reduction, pattern identification, compression of text	Social group analysis by means of clustering, processing the context awareness, managing the user profiles, and predicting the mobility
Machine learning	Classification, regression, neural network algorithm	Identifying the context, predicting the traffic, identifying the user location
	Deep learning	Complex wireless data processing
	Reduction algorithms	Extracting the data using feature extraction, compression techniques for data usage, storage
	Decomposition—primal/dual, Archived Data Manager	Routing distribution and controlling the rate of transfer, resource allocation

such as clustering, are used in mobile computing contexts as a technique whereby the movement of the user is sensed with the sensors connected to wireless networks (10).

3. **ML:** The core aim of these algorithms is to find the efficient relationship between the input and output data with corresponding actions. It leads to finding the hidden patterns in data. Classification and regression are the most commonly used algorithms in wireless network, where they are applied to identify the wireless usage context and support in forecasting traffic flow. It is also used to identify the location of users and their movements, which assists them in real-time actions to be taken based on traffic flow. It also uses deep learning, reduction, and decomposition algorithms in applications such as wireless data processes in complex networks, extracting data using feature extraction techniques, and the compression in data storage and usage. For traffic-routing applications, ML can be used for distributing and controlling the flow and resource allocation of wireless networks:

   - Distributed optimization algorithms, like primal/dual decomposition and interchanging the multiplier direction, are helpful in decoupling large-scale problems in statistics; the main problems are divided into subproblems for the purpose of parallel computing, which reduces the central processing time by reducing the bandwidth load.
   - Dimension reduction techniques are used for reducing the volume of data in big data processing. Different techniques are used to analyze key components that have a relationship with various factors, which has been adopted recently.
   - Other learning models used in wireless network analysis recently are handling complex real-time accessing data at varying time intervals used to take decisions periodically based on statistical models along with deep learning techniques.

5.4 CHALLENGES AND BENEFITS OF WIRELESS NETWORKS

Everything in this world comes with significant challenges, so employing the analytics of big data for controlling and optimizing the wireless network also faces the same challenges (39–42). Distinctive challenges are posed in the system of administrating and influencing the gargantuan number of facts; algorithm formulation for compelling and competent processing of various data sets has been utilized to gain insights from that information in network analysis. The extent of efforts, the workforce needed to handle and promote the big data platform, and the scope of endeavor skills are primary to the prominent challenging tasks faced by operators in wireless communication as is losing information in their direct control. This occurs due to the automation and concurrent operations happening in the structure of big data analytical network. By considering the future of wireless communication, this concern is unavoidable and assertive. Above this, a huge stake is also needed in studying this.

Operators of wireless communication view these challenges to the benefits by the analysis of facts in big data through its efficiency in offering the services throughout the network. This, in turn, helps providers do better groundwork; through the high use of resources in wireless communication and maintenance and reducing the cost of operation, it also facilitates providers' flexibility in offering their services based on users' requirements. With the advancement in technology analysis and more gadgets, providers are able to fetch users' requirements based on these various analyses, and it effectively manages traffic on wireless networks, balancing networks equally among various resources.

5.5 CHALLENGES IN DATA ACQUISITION

Following are the challenges faced in the acquisition of data:

- Representation Difficulty: As we know, information is acquired through various sources and forms, such as in structured and unstructured ways. In case of mobile networks, we will have data from both the system and end users in various formats. The challenge here is determining how to exhibit these various formats in big data analytics in wireless communication.
- Efficiency in Data Collection: Procuring various sources of raw data is referred as data collection. The procedure for this should be precise and justifiable as inaccuracy leads to a fault in any analysis of network communication.
- Data Transmission Efficiency: The challenge arises in the way the data are transmitted because of the huge volume involved. The reasons for this challenge can be due to (1) its high bandwidth usage and (2) the efficient consumption of energy.

5.6 OPPORTUNITIES IN DATA ACQUISITION

In these challenges, we are able to identify some opportunities in data acquisition toward research:

- As we are aware, data are acquired through various formats, which is contradictory to operational data, which is in an organized format (37). Currently, no tool or method is available to handle these heterogeneity data. This ensures obstacles in the preprocessing, storage, and analysis of data. A new trend in research is rising in the handling of heterogeneous data types.
- Data transmission security: while data are transferred from various wireless gadgets, there is a possibility of malicious attacks due to the interference in wireless devices. Let us take the example of a narrow-band IoT security structure that has been taken away to minimize the cost (20). Current research has proved the efforts in secure secure-key-generation-based reciprocity and randomness in wireless communication, and protection from jamming (28, 29) has been explained with the IoT.
- Regenerating energy in data transmission: adding to security concerns during transmission from wireless devices, regenerating energy is also a challenge. New energy regenerator techniques, such as radio frequency (RF)–enabled, provides a remedy by connecting RFID tags by air. To control the greater diminishing of RF energy, the transfer of wireless power is expected in a directional way (21).

5.7 CHALLENGES IN DATA PREPROCESSING

The data acquired from wireless communication have the following benefits while preprocessing those data:

- Various types of data are sensed by different gadgets.
- The inaccuracy of data is due to a loss of communication, a failure of networks, or intervention during the collection of data (22).
- The duplication of data occurs more as many users read the same data at different intervals (23). This leads to variability in the data. In the health care system, this wireless technology has created numerous duplications of medical information (24). Database reliability improves based on redundancy of the data, but duplication produces an unpredictable result, particularly in preprocessing. These benefits lead to the challenges in preprocessing such as the following:
- Integrating various data types: as we mentioned, TR various types of data, along with a variety of features, are generated through wireless networks. To implement an efficient big data analysis, these various sources and types must be integrated to provide efficient analytics (38).
- Reducing the duplication of data gives rise to an inconsistency in the data that makes to do consequent data analysis.
- Cleaning and compressing data: due to the vast size of data in big data analytics, cleaning data becomes a tedious task. An effective method has to be implemented to clean the data without any fault in data.

5.8 OPPORTUNITIES IN DATA PREPROCESSING

- Privacy in data preprocessing has to be considered as many studies state that during preprocessing, there is a lack the concern for users' personal information (25). How to complete preprocessing data without affecting users' privacy is a big question.
- Assuring security during preprocessing in large-scale data analytics is crucial as the analyses are done in different phases, providing opportunities for security concerns. Studies say that IoT systems are correspondingly facing problems in updating data due to malicious attacks during transfer (26).

5.9 CHALLENGES IN DATA STORAGE

The storage of data plays a vital role in analysis. In the case of large-scale wireless communication, challenges arise in designing data storage in an efficient manner. The following are a few challenges that can be seen during the storage of data:

- Consistent and uninterrupted data storage: it is difficult to maintain consistency and uninterrupted data storage due to the consideration of the cost of storing a huge volume of data (37).

- Other than storage consistency, another challenge is the expandability of storage systems in the analysis of big data. The various types and huge volume of data in wireless communication lead the traditional storage being impractical for use in analyses.
- Another concern in data storage is efficiency. To support the huge volume of simultaneous access of data, the analysis needs data storage efficiency along with consistency and extendibility.

5.10 OPPORTUNITIES IN DATA STORAGE

The various research opportunities in data storage are as follows:

- Few studies have been conducted regarding various types of wireless networks related to distributive models and their storage. Depending on the type of wireless communication, data storage and its model are needed.
- An airy computing model is needed to store data as the assumption is that data are stored either in mobile gadgets or in cloud storage, as many wireless devices have restrictions on their smaller storage and interpretation.
- Assurances of security and privacy are also a wide area to study in wireless communication. Different people have suggested numerous privacy methods for wireless technologies without losing the originality of the data.

5.11 CONCLUSION

In this chapter, we have tried to give an overview of ML approaches in big data analytics optimization for wireless sensor networks, where it helps network operators to use big data analytics, ML, and AI approaches in networks to control and operate network optimization. In this chapter, we also discussed the various drivers used in analytics and the types of analytics. The role of big data analytics, ML and AI in wireless communication, and the various optimization for the data driven from various network perspectives were elaborated on. Finally, we discussed some of the opportunities and issues faced by wireless network services.

REFERENCES

1. M. Paolini. *Mastering Analytics: How to Benefit From Big Data and Network Complexity*, https://www.iotforall.com/analytics-big-data-network-complexity https://www.prweb.com/releases/2017/06/prweb14470349.htm. Accessed 2 November 2017.
2. Dai, Hong-Ning, Hao Wang, Raymond Wong, and Zibin Zheng. (2019). Big Data Analytics for Large Scale Wireless Networks: Challenges and Opportunities. *ACM Computing Surveys*. (accepted to appear), doi:10.1145/3337065.
3. Kibria, Mirza et al. (2018). Big Data Analytics, Machine Learning and Artificial Intelligence in Next-Generation Wireless Networks. *IEEE Access*, PP. doi:10.1109/ACCESS.2018.2837692.
4. Bi, S., R. Zhang, Z. Ding, and S. Cui. (2015). Wireless Communications in the Era of Big Data. *IEEE Communications Magazine*, 53(10), 190–199, October.

5. Montori, F., L. Bedogni, and L. Bononi. (2018). A Collaborative Internet of Things Architecture for Smart Cities and Environmental Monitoring. *IEEE Internet of Things Journal*, 5(2), 592–605, April.

6. Leng, K., L. Jin, W. Shi, and I. Van Nieuwenhuyse. (2018). Research on Agricultural Products Supply Chain Inspection System Based on Internet of Things. *Cluster Computing*, 22, February.

7. Dweekat, A. J., G. Hwang, and J. Park. (2017). A Supply Chain Performance Measurement Approach Using the Internet of Things: Toward More Practical SCPMS. *Industrial Management & Data Systems*, 117(2), 267–286.

8. Zhong, R. Y., C. Xu, C. Chen, and G. Q. Huang. (2017). Big Data Analytics for Physical Internet-Based Intelligent Manufacturing Shop Floors. *International Journal of Production Research*, 55(9), 2610–2621.

9. Hristova, D., M. J. Williams, M. Musolesi, P. Panzarasa, and C. Mascolo. (2016). Measuring Urban Social Diversity Using Interconnected Geo-Social Networks. *Proceedings of the 25th International Conference on World Wide Web (WWW). ACM*, 21–30.

10. Krause, A., A. Smailagic, and D. P. Siewiorek. (2006). Context-Aware Mobile Computing: Learning Context-Dependent Personal Preferences from a Wearable Sensor Array. *IEEE Transactions on Mobile Computing*, 5(2), 113–127, February.

11. Cheng, X., L. Fang, L. Yang, and S. Cui. (2017). Mobile Big Data: The Fuel for Data-Driven Wireless. *IEEE Internet Things of Journal*, 4(5), 1489–1516, October.

12. Cheng, X., L. Fang, X. Hong, and L. Yang. (2017). Exploiting Mobile Big Data: Sources, Features, and Applications. *IEEE Networks*, 31(1), 72–79, January–February.

13. Jiang, C., H. Zhang, Y. Ren, Z. Han, K. C. Chen, and L. Hanzo. (2017). Machine Learning Paradigms for Next-Generation Wireless Networks. *IEEE Wireless Communications*, 24(2), 98–105, April.

14. Kyriazakos, S. A., and G. T. Karetsos. (2004). *Practical Radio Resource Management in Wireless Systems*. Boston, MA: Artech House.

15. Procera Networks. *RAN Perspectives: RAN Analytics & Enforcement*, www.proceranetworks.com/hubfs/Resource%20Downloads/Datasheets/Procera_DS_RAN_Perspectives.pdf?t=1481193315415. Accessed 13 October 2017.

16. Zhou, Xuan et al. (2013). Human Mobility Patterns in Cellular Networks. *Communications Letters, IEEE*, 17, doi:10/1877–1880.10.1109/LCOMM.2013.090213.130924.

17. Banerjee, A. (2014). *Advanced Predictive Network Analytics: Optimize Your Network Investments & Transform Customer Experience*. New York: Heavy Reading, White Paper, February.

18. Jung, Byoung, Nah-Oak Song, and Dan Sung. (2013). A Network-Assisted User-Centric WiFi-Offloading Model for Maximizing Per-User Throughput in a Heterogeneous Network. *IEEE Transactions on Vehicular Technology*, 63.

19. Hafiz, H., H. Aulakh, and K. Raahemifar. (2013). Antenna Placement Optimization for Cellular Networks. *2013 26th IEEE Canadian Conference on Electrical and Computer Engineering (CCECE), Regina, SK*, 1–6, doi:10.1109/CCECE.2013.6567765.

20. Xu, J., J. Yao, L. Wang, Z. Ming, K. Wu, and L. Chen. (2017). Narrowband Internet of Things: Evolutions, Technologies and Open Issues. *IEEE Internet of Things Journal*, PP(99), 1–13.

21. Bi, S., C. K. Ho, and R. Zhang. (2015). Wireless Powered Communication: Opportunities and Challenges. *IEEE Communications Magazine*, 53(4), 117–125.

22. Li, M., D. Ganesan, and P. Shenoy. (2009). Presto: Feedback-Driven Data Management in Sensor Networks. *IEEE/ACM Transactions on Networking*, 17(4), 1256–1269, August.

23. Ertek, G., X. Chi, and A. N. Zhang. (2017). A Framework for Mining RFID Data from Schedule-Based Systems. *IEEE Transactions on Systems, Man, and Cybernetics: Systems*, 47(11), 2967–2984.

24. Zhang, Y., M. Qiu, C. W. Tsai, M. M. Hassan, and A. Alamri. (2017). Health-CPS: Healthcare Cyber-Physical System Assisted by Cloud and Big Data. *IEEE Systems Journal*, 11(1), 88–95.

25. Wang, N., X. Xiao, Y. Yang, T. D. Hoang, H. Shin, J. Shin, and G. Yu. (2018). PrivTrie: Effective Frequent Term Discovery Under Local Differential Privacy. *IEEE International Conference on Data Engineering (ICDE) IEEE*, 630–649.

26. Roman, R., J. Zhou, and J. Lopez. (2013). On the Features and Challenges of Security and Privacy in Distributed Internet of Things. *Computer Networks*, 57(10), 2266–2279, July.

27. Xu, W., S. Jha, and W. Hu. (2019). Lora-Key: Secure Key Generation System for Lora-Based Network. *IEEE Internet of Things Journal*, 1–10. (early access).

28. Balachandar, S., and R. Chinnaiyan. (2018). Centralized Reliability and Security Management of Data in Internet of Things (IoT) with Rule Builder. *Lecture Notes on Data Engineering and Communications Technologies*, 15, 193–201.

29. Hu, L., H. Wen, B. Wu, F. Pan, R. Liao, H. Song, J. Tang, and X. Wang. (2018). Cooperative Jamming for Physical Layer Security Enhancement in Internet of Things. *IEEE Internet of Things Journal*, 5(1), 219–228, February.

30. Balachandar, S., and R. Chinnaiyan. (2018). Reliable Digital Twin for Connected Footballer. *Lecture Notes on Data Engineering and Communications Technologies*, 15, 185–191.

31. Balachandar, S., and R. Chinnaiyan. (2018). A Reliable Troubleshooting Model for IoT Devices with Sensors and Voice Based Chatbot Application. *International Journal for Research in Applied Science & Engineering Technology*, 6(2), 1406–1409.

32. Swarnamugi, M., and R. Chinnaiyan. (2018). IoT Hybrid Computing Model for Intelligent Transportation System (ITS). *IEEE Second International Conference on Computing Methodologies and Communication (ICCMC)*, 15–16, February.

33. Swarnamugi, M., and R. Chinnaiyan. (2017). Cloud and Fog Computing Models for Internet of Things. *International Journal for Research in Applied Science & Engineering Technology*, 2, December.

34. Sabarmathi, G., and R Chinnaiyan. (2019). Envisagation and Analysis of Mosquito Borne Fevers: A Health Monitoring System by Envisagative Computing Using Big Data Analytics. *Lecture Notes on Data Engineering and Communications Technologies (LNDECT), Book Series.* Cham: Springer, vol. 31, 630–636.

35. Balachandar, S., and R. Chinnaiyan. (2019). Internet of Things Based Reliable Real-Time Disease Monitoring of Poultry Farming Imagery Analytics. *Lecture Notes on Data Engineering and Communications Technologies (LNDECT), Book Series.* Cham: Springer, vol. 31, 615–620.

36. Swarnamugi, M., and R. Chinnaiyan. (2019). IoT Hybrid Computing Model for Intelligent Transportation System (ITS). *Proceedings of the Second International Conference on Computing Methodologies and Communication (ICCMC 2018)*, 802–806.

37. Sabarmathi, G., and R. Chinnaiyan. (2016). Big Data Analytics Research Opportunities and Challenges—A Review. *International Journal of Advanced Research in Computer Science and Software Engineering*, 6(10), 227–231.

38. Sabarmathi, G., and R. Chinnaiyan. (2020). Investigations on Big Data Features Research Challenges and Applications. *IEEE Xplore Digital Library International Conference on Intelligent Computing and Control Systems (ICICCS)*, 782–786.

39. Divya, R., and R. Chinnaiyan. (2017). Reliability Evaluation of Wireless Sensor Networks (REWSN—Reliability Evaluation of Wireless Sensor Network). *2017 International Conference on Intelligent Computing and Control Systems (ICICCS), Madurai*, 847–852, doi:10.1109/ICCONS.2017.8250583.

40. Divya, R., and R. Chinnaiyan. (2018). Reliable Smart Earplug Sensors for Monitoring Human Organs Based on 5G Technology. *2018 Second International Conference on Inventive Communication and Computational Technologies (ICICCT), Coimbatore*, 687–690, doi:10.1109/ICICCT.2018.8473218.

41. Divya, R., and R. Chinnaiyan. (2019). Reliable AI-Based Smart Sensors for Managing Irrigation Resources in Agriculture—A Review. In S. Smys, R. Bestak, J. Z. Chen, and I. Kotuliak. (eds), *International Conference on Computer Networks and Communication Technologies*. Lecture Notes on Data Engineering and Communications Technologies. Singapore: Springer, vol. 15, doi:10.1007/978-981-10-8681-6_25.

42. Divya, R., and R. Chinnaiyan. (2019). Detection of Objects and Evaluation with the IO Link Using Miniature Laser Sensor—A Review. In J. Hemanth, X. Fernando, P. Lafata, and Z. Baig. (eds), *International Conference on Intelligent Data Communication Technologies and Internet of Things (ICICI) 2018*. ICICI 2018, Lecture Notes on Data Engineering and Communications Technologies. Cham: Springer, vol. 26, doi:10.1007/978-3-030-03146-6_83.

6 Improved Video Steganography for Secured Communication Using Clustering and Chaotic Mapping

Lakshmi Harika Palivela, M.R. Sumalatha,
A. Naveen, S. Maithreyan, and S.V. Vijay

CONTENTS

DOI: 10.1201/9781003107477-6

6.1 INTRODUCTION

With emerging technologies, information can be easily distributed and transmitted across the network through various protocols. However, in this data communication and transmission, protection becomes a critical concern. The data have to be secured from unauthorized users or mischievous attackers, and an efficient model is needed. Steganography proves to be an efficient and secure way of transferring sensitive data like text, image or video. The type of steganography used for concealing the information depends on the medium used to cover the secret information. If an image is used as a wrap for hiding the sensitive information, then it is called image steganography. If the cover is a video, then it is called video steganography. Although both types of steganography techniques prove to be secure, the type of steganography can be chosen based on the application. Steganography in combination with a cryptography model may achieve better security for the hidden message. Steganalysis is the study of detecting messages hidden using steganography (breaking); this is analogous to cryptanalysis. In various applications, like secret data storing and confidential data communication, we make use of steganography. A further increase insecurity can be achieved by applying encryption algorithms, shuffling of blocks or chaotic maps to create confusion in the output stego-video combined with enhancement algorithms for de-noising and sharpening for improving the output quality.

6.2 RELATED WORKS

In a recent survey, many works are based on secure multimedia communication via wireless networks. An improved version for enforcing data security in wireless communication is steganography (1–19).

The term steganography in combination with a cryptography model may achieve better security for the hidden message (3–5). Jie Wang et al. (6) proposed high efficiency video coding steganography model based on intra prediction mode (IPM) approach. In their concept, a covering rule is suggested on the videos in conjunction with the coding information, which can boost the protection efficiency of a masked video stream. In addition to preserving the video quality and security efficiency, the implementation is very easy. However, it leads to a bit rate increase, which is a disadvantage of this approach.

Ramathan et al. (7) proposed a reliable steganographic model in combination with discrete wavelet and cosine transform on MOT (multiple object tracking). The secret message is preprocessed by applying Hamming codes to encrypt the secret information. Next, a motion-based model is used for host videos to identify the interested regions on the frames. Then the method of hiding the data is carried out. The embedding rate is high in this method; our proposed approach could achieve better results with a low embedding rate.

Yingnan Zhang et al. (8) propose a secret sharing model by integrating grey relational analysis with the video compression format partition mode H.264/AVC. The algorithm is reliable and a good ability of stegoanalysis and a slight effect on the video carrier bit rate. The rate of bit error in the simulation environment is small, but it results in frame loss due to noise. Zhuo Zhang et al. (9) propose a steganography by cover synthesis approach to perform steganography, in particular on picture frames, using generative adversarial network adversarial network to make encrypted pictures from secret messages. This generative model shows the relationship between the images and hidden messages with labels that can be given to a classifier network. This is completely based on an image data set.

Elshazly Emad et al. (10) propose a secure model to cover bit streams of the confidential text message with the less important bits of wavelet transform of grey-scale images. It extracts the hidden data efficiently without using the original wrapper image. As compared to this method, our proposed approach achieved a much better result for peak signal-to-noise ratio (PSNR). Donghui Hu et al. (11) used a qualified neural network model to map hidden data into a noise vector. This method results in extracting precise information. Our proposed method overcomes the limitations of all kinds of steganalysis as block shuffling is used. Kaixi Wang et al.'s (12) work used a secure communication model that stands for Chinese character stroke numbers with a low embedding rate. Thus, a secret message is indexed to a list of uniform resource locators for secure communication.

Ayesha Saeed et al. (13) presented a content-adaptive steganography approach aimed at defining the nature of the image's texture material for data masking. The density of the material is evaluated with high-pass filters that measure the association of pixels in eight directions. The pixels are then grouped from the highest to the lowest in priority of complexity. Xinghong Qin et al. (14) propose a steganographic model on color images by modelling pixels of color at a spatial location. Guo et al. (15) propose simple image-based encryption technique to perform the confusion and diffusion of input plain image. Shan-Shan Li et al. (16) propose steganographic model using density-based spatial clustering of applications with noise (DBSCAN) by constructing a cover tree for retrieving neighbor points on the covered video. Our proposed approach can be employed on large-scale data. To overcome the limitations, the proposed model involves using a density-based clustering technique like DBSCAN to improve PSNR, because the addition of extra information causes distortion of the actual video in steganography, and to improve the embedding capacity, and block shuffling adds more resistance toward steganalysis attacks and protects the secret from the intruder. The chaotic mapping technique is used to shuffle the blocks, thereby creating chaos in the stego-video.

6.3 DATA SECURITY TECHNIQUES

6.3.1 CRYPTOGRAPHY

Cryptography involves the encryption of information using various substitution and rearrangement ciphers. Only the targeted receiver can get the actual data upon decryption using the required private keys. In cryptography, the original readable text or secret information is converted into an unreadable text called a cipher text. The cipher text can be known to the intruder or third party before the decryption stage. Although cipher text is prone to malicious attacks, cryptography provides a basic level of security in wireless data communication in Figure 6.1.

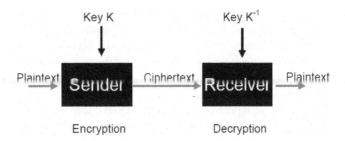

FIGURE 6.1   Cryptography.

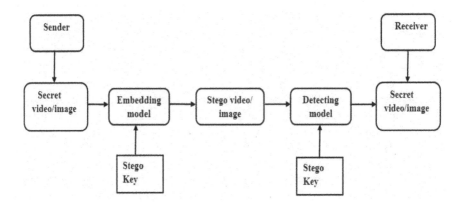

FIGURE 6.2   Steganography.

6.3.2 STEGANOGRAPHY

Steganography is the art of masking information or sensitive data inside an image or video. In image steganography, we use a cover image to hide the secret, which can be an image or text, whereas in video-based steganography, a video is used for covering the sensitive information. In steganography, the intruder is unaware of the fact that something is hidden within the carrier medium. As long as the presence of the secret remains unknown to the third party or an intruder, the possibility of attacks on the stego-image or stego-video remains lower. This asset adds some extra security to the steganography in Figure 6.2.

6.3.3 VIDEO STEGANOGRAPHY

Various types of steganography methods are available depending on the type of cover medium used. A video is used to mask the payload in video steganography. A video is a stream of frames, and each frame is similar to a still image. Video is generated by the sequential arrangement of frames. Because the dimensions of the video are greater than the other digital cover media, it is possible to cover secret information within a video.

6.3.3.1 Research Challenges

The challenges in the video steganography have been listed below:

The human visual system will not notice the small changes that occur in the cover data. However, the steganalysis tools are more likely to break the weak steganographic techniques to protect the data; researchers have to develop strong steganography techniques. The success of a steganography technique relies on important factors like the robustness against attacks, the capacity of embedding and imperceptibility.

The term *embedding ability* is the sum of payload that can be carried inside a video or picture cover. Stego's visual consistency, stability and resistance to attackers are involved in incorporating performance. High embedding performance will reduce the assumption of the intruder seeking secret information, making it more difficult to locate it through steganalysis tools.

The imperceptibility of a steganographic algorithm is also a potential challenge since the payload has to be encoded into the cover image or video. This steganographic method is imperceptible because the human eye is unable to differentiate between the covering media and the embedded or stego-image. Higher imperceptibility improves the performance of steganography as the human visual system cannot identify the encoding of data within the cover image or video which reduces attacker's suspicion. Although steganography is considered more secure than cryptography, both of them are combined together to make the data more robust against attacks.

6.4 METHODOLOGY

6.4.1 Implementation of Video Steganography with Clustering and Chaotic Mapping for Secure Communication

At the sender end point, the video is used as a cover medium to encrypt the frames of secret video. In wireless sensor networks (WSNs), to enhance the secure communication in multimedia, that is, stego-image or stego-video, we can further use encryption algorithms like block shuffling or chaotic mapping, which adds extra security to the hidden information. Both imperceptibility and robustness should be satisfied. Overhead can be a main issue in steganography, but the security of data in wireless communication is a major concern. Overhead can be reduced by optimizing the key frame extraction technique and optimized clustering techniques.

6.4.2 Proposed System for Secure Communication

There are two stages involving encryption and decryption. The encoding stage initially involves extracting the key frames from the secret or hidden video to be transferred securely through the network. For key frames extraction, we use the Euclidean distance method. The DBSCAN clustering is done on cover video to increase PSNR. Later, the trained model is used to compress the information from the secret video frames into the least noticeable portions of the cover video frames. Later, block shuffling by a cat mapping technique is used for confusion (chaos) and increases the security of the stego-video. In the decoding stage, these processes are performed in the reverse order. The system architecture is shown in Figure 6.3.

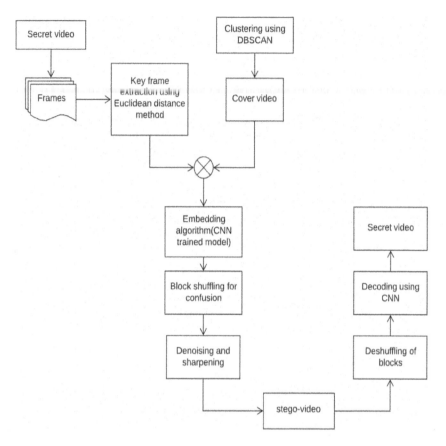

FIGURE 6.3   Encoding and Decoding Model for Secure Communication.

6.4.2.1 Extraction of Key Frames

Key frames are considered as a frame encoded without reference to any image in another frame for compression problems, and these key frames are referred to as intra-coded frames, or I frames. These key frames provide a compendious representation of the given video series that reflects the slow and quick motion of the selected shot. Using the Euclidean distance method, the key frames are extracted from the secret video. The edges of the image are initially found in each frame in the Euclidean distance method, and the edge sum is calculated for the frames. If the edge sum of the difference is greater than the T threshold, the key frame count will be increased by also considering the current frame as a key frame.

6.4.2.2 Clustering of Cover Video Frames

The multilevel clustering enhances the performance of the steganography algorithm in WSNs. Here, frames of the cover media are clustered so that the secret video frames can be easily hidden and retrieved more effectively while comparing with usage of a cover medium without clustering. Clustering has been proved to increase the performance of video steganography by increasing its PSNR and decreasing the mean squared error (MSE) term. DBSCAN clustering is done on the cover video, and the resultant GIF file is stored. This clustered video can now serve as a cover medium for secret data. For performing encoding and decoding the video in the training, the convolutional neural network (CNN) network is used to process the cover and secret media and embed the hidden video within the cover media. The steganography algorithm for secure wireless communication is shown after Algorithm I.

Algorithm I: Improved Video Steganography for Secure Communication

Input parameter: Video frames {A series of images} f_i = 1, 2, 3, . . . , Ni (Ni = total no. of frames)

Output parameter: Frame reduced video that is compressed in size.

1. Select video clip.
2. Select first video frame as the initial key frame.
3. Apply grey quantization and edge detection to find edges of the image in the first frame.
4. Find the sum of the edges of the first (previous) frame.
5. The specific function of edge sum is fixed as threshold (T).
6. For each frame, f = 1,2,3, . . . , N.
    6.1. Apply the edge detection algorithm; find the edges and calculate the edge sum.
    6.2. Find edge sum difference between the previous and the current frame.
    6.3. If percentage T < diff. sum
        6.3.1. Store the previous selected key-frame in buffer as a real Key Frame
        6.3.2. Put the current selected key frame in buffer as a keyframe (f).
7. Display the key frames in the form of compressed video.
8. Performing clustering to the cover video
    8.1 Set cluster index C
        8.1.1. For each frame f in video
                For each pixel P, P′ in f
                    Set point P as visited.
                    Ni = neighbours of P
                    if |Ni| >= minpts then
                        Ni = Ni U Ni′
                    if pixel P′ not in C
                    then
                        C+ = P′
9. Display the clustered frames.
10. Embed the secret video (sv) and cover video (cv).
    10.1. CNN is used to train the hidden network with input
                Concatenate (sv,cv)
                Display output (sgv)   //stego video

11. Apply chaotic mapping.
    11.1. Set iterations N to 10
        For each frame f in video
            Read image pixels X_i and Y_i
            For 1 to N
            For X to N
            For Y to N
          Perform cat mapping by
              $(xi,yi) = (2X_i + Y_i, X_i + Y_i) \bmod 1$
    11.1.2 Store new pixel values in array
12. Return encrypted frames.
13. For each frame f in the revealing network
    Decode (sgv)-> Output(sv)
14. Receiver receives the original Video

6.4.2.3 Block Shuffling Using Chaotic Mapping

The main aim of steganography and cryptography algorithms is to maintain information secrecy. Also, in image encryption techniques, to enforce more security, chaotic mapping techniques are applied to protect the data from third parties. The same techniques can also be extended to the videos since they are a collection of frames or images. One such chaotic mapping technique is the Arnold cat map, which is applied on the stego-video.

Arnold cat map is a chaotic mapping technique for encrypting the frames or images by randomly permuting the pixels in a frame. The concept of the algorithm is to continuously rotate the image so that it becomes a form that is not visible and random so that the image cannot be seen by the naked eye, making it less suspicious to attacks.

Three discrete two-dimensional chaotic invertible maps techniques are baker map, standard map and Arnold cat map, which are commonly used for performing permutation. Such map techniques generate injection mapping that transfers every pixel in the image to a different location. Usually, this step is typically taken many times to achieve a proper effect of uncertainty. When compared to other maps, the Arnold cat and baker maps possess short time downside after discretion and are able to finish the permutation in a short time without any complications in floating values. Because of its lowest computational complexity, the Arnold cat map technique is used for image/video frame permutation in the current analysis. The mathematical formula of a chaotic map of Arnold is stated as follows:

$$
\begin{bmatrix} X_{n+1} \\ Y_{n+1} \end{bmatrix} = \begin{bmatrix} 1 & 1 \\ 1 & 2 \end{bmatrix} \begin{bmatrix} X_n \\ Y_n \end{bmatrix} \bmod 1 = A \begin{bmatrix} X_n \\ Y_n \end{bmatrix} \bmod 1. \tag{6.1}
$$

By adding two control parameters 'p' and 'q', the cat map can be simplified:

$$
\begin{bmatrix} X_{n+1} \\ X_{n+1} \end{bmatrix} = \begin{bmatrix} 1 & p \\ q & pq+1 \end{bmatrix} \begin{bmatrix} X_n \\ Y_n \end{bmatrix} \bmod 1 = A \begin{bmatrix} X_n \\ Y_n \end{bmatrix} \bmod 1. \tag{6.2}
$$

Since 'X' and 'Y' numbers are real, however, the generic mapping technique is not suitable for the image/video frame permutation that operates on a finite number of lattice pixels. The generalized map therefore needs to be discretized so that it can be integrated into the cryptosystem to perform the permutation among pixels.

The discrete cat map is formulated by altering the range value (X, Y) from a square unit by adjusting the lattice:

$$
\begin{bmatrix} X_{n+1} \\ Y_{n+1} \end{bmatrix} = \begin{bmatrix} 1 & p \\ q & pq+1 \end{bmatrix} \begin{bmatrix} X_n \\ Y_n \end{bmatrix} \bmod Ni = A \begin{bmatrix} X_n \\ Y_n \end{bmatrix} \bmod Ni, \tag{6.3}
$$

where 'Ni' is a square image, height or width and variables 'p or q' are the permutation key.

6.5 RESULT ANALYSIS

The experimental analysis is carried out using Python and its performance is estimated by embedding secret video within various cover videos like Foreman_qcif. yuv, Stefan_qcif.yuv and so on. The clustering makes the pixels of an image to be distributed un-uniform manner which can be visualized using histograms. The random spikes in the histogram represent the group of pixels in an image that are clustered together based on the clustering algorithm. As we can infer from the histogram of the clustered frames, DBSCAN clustering proves to be more effective compared with the k-means clustering since the histogram shows more clusters, which are visualized as peaks in DBSCAN when compared with k-means.

The histograms comparison for a sample Stefan_qcif.yuv video after clustering is shown in Figure 6.4, Container_qcif.yuv in Figure 6.5 and Forman_qcif.yuv in Figures 6.6 and 6.7.

The analysis of stego-video frames of four videos is done by evaluating the following parameters.

6.5.1 RELATIVE PEAK SIGNAL-TO-NOISE RATIO

To check the resemblance among two picture frames and specify how they are identical to each other is given by PSNR, and in contrast, MSE shows the dissimilarity among two images. PSNR is calculated as follows:

$$
\text{PSNR} = 10 \log 10(I2/\text{MSE}). \tag{6.4}
$$

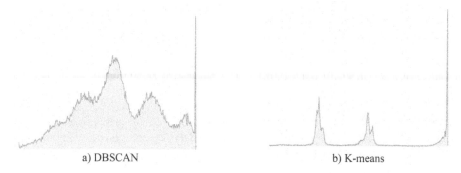

a) DBSCAN   b) K-means

FIGURE 6.4   Histograms of Clustered Video Frame of Stefan_qcif.yuv.

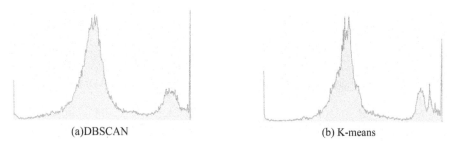

(a)DBSCAN   (b) K-means

FIGURE 6.5   Histograms of Clustered Video Frame of Container_qcif.yuv.

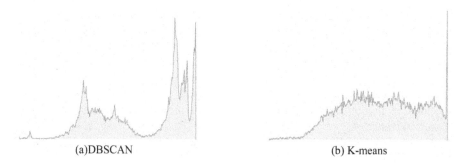

(a)DBSCAN   (b) K-means

FIGURE 6.6   Histograms of Clustered Video Frame of Forman_qcif.yuv.

For a given picture, the PSNR value means nothing, which is used to measure its quality. As a result, we separated the PSNR of each reference (ref) image or video frame by the PSNR of its corresponding stego-image or video frame. However, the reference is a compressed clean image of the source, while its PSNR (ref) is calculated

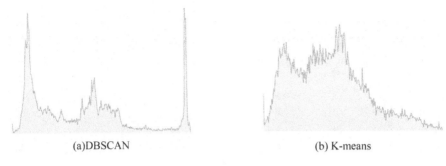

(a)DBSCAN (b) K-means

FIGURE 6.7 Histograms of Clustered Video Frame of Carphone_qcif.yuv.

against the uncompressed original source image, and it estimates the quality among the pictures at each point. It is calculated using the following formula:

$$PSNR_r = PSNR_{ref}/PSNR_{stego}. \tag{6.5}$$

6.5.2 Pixel Change Rate Number

The number of pixel change rate (NPCR) checks whether a single pixel in original source image can produce adequate diffusion in the cipher image or video frame. Let two cipher images with a one-pixel difference be $A(i, j)$ and $A'(i, j)$. $K(i, j)$ is similar matrix to store the results between $A(i, j)$ and $A'(i, j)$. So, if $A(i,j) = A'(i, j)$, then $K(i, j) = 1$; otherwise, $K(i, j) = 0$. The NPCR shall be

$$NPCR = \frac{\sum_{i,j} K_{(i,j)}}{a*b} *100\%. \tag{6.6}$$

6.5.3 Unified Average Changing Intensity

The percentage difference between two pixels is determined by unified average changing intensity (UACI). For pixels $A(i, j)$ and $A'(i, j)$ picture frame, it is expressed as

$$UACI = \frac{\left\{\sum_{i,j} \dfrac{\left|A_{(i,j)} - A'_{(i,j)}\right|}{225}\right\}}{a*b} *100\%. \tag{6.7}$$

6.5.4 MSE

MSE is a distortion measure. MSE value computes the variations between the image or video frame being processed (A_{ij}) and its reference image frame (B_{ij}). The lower the MSE value, the better the result. It is expressed as shown in Table 6.1:

TABLE 6.1

Statistical MSE Results for Sample Cover Video

Video Sequence	Number of Frame	MSE (Ideal value = 0)		NCC (Ideal value = 1)		NAE (Ideal value = 0)	
		k-Means	DBSCAN	k-Means	DBSCAN	k-Means	DBSCAN
Container_qcif.yuv	300	0.2569	0.1755	0.7982	0.8642	0.0912	0.0829
Foreman_qcif.yuv	300	0.6506	0.2160	0.7316	0.8396	0.0819	0.0732
Carphone_qcif.yuv	382	0.4751	0.2336	0.5430	0.7312	0.0532	0.0481
Stefan_qcif.yuv	90	0.8247	0.4323	0.4932	0.6401	0.0682	0.0585

$$MSE = \frac{1}{mn}\sum_{i=1}^{m}\sum_{j=1}^{n}\left(A_{ij} - B_{ij}\right)^{2}. \tag{6.8}$$

6.5.5 Normalized Cross-Correlation

It shows the comparison between the processed (A_{ij}) and reference image (B_{ij}). It is expressed as

$$NCC = \sum_{i=1}^{m}\sum_{j=1}^{n}\frac{\left(A_{ij} \times B_{ij}\right)}{A_{ij}^{2}}. \tag{6.9}$$

6.5.6 Normalized Absolute Error

A higher value of normalized absolute error (NAE) represents an image of poor quality. It is expressed as

$$NAE = \sum_{i=1}^{m}\sum_{j=1}^{n}\frac{\left(\left|A_{ij} - B_{ij}\right|\right)}{A_{ij}}. \tag{6.10}$$

The experimental analysis carried out in the python environment, and its performance is computed by embedding a secret video within various cover videos (19). The video is sampled at 30 frames per second with resolution 176*144. The sample results obtained during the process of execution are shown in the following figures.

The sample results of the key frames in the secret video shown in Figure 6.8. The video considered here is MISSILE (18) video having frames 689 in total.

The sample results of the proposed model shown in Figures 6.9 and 6.10 both the source or sender side and the destination or receiver side to cover the hidden media for secure communication as shown in the following sections.

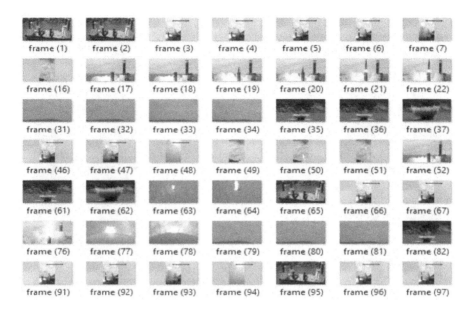

FIGURE 6.8 Extraction of Key Frames in Secret Video.

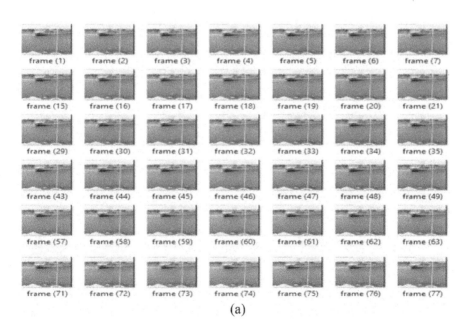

(a)

FIGURE 6.9 For Video Container_qcif.yuv: (a) Frame Conversion (b) Stego-Video after Chaotic Mapping (encrypted video frames).

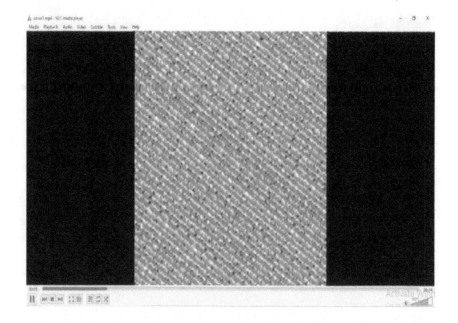

(b)

FIGURE 6.9   (Continued)

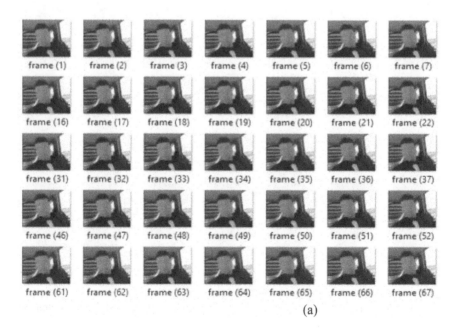

(a)

FIGURE 6.10   For Video Foreman_qcif.yuv: (a) Frame Conversion (b) Stego-Video after Chaotic Mapping (encrypted video frames).

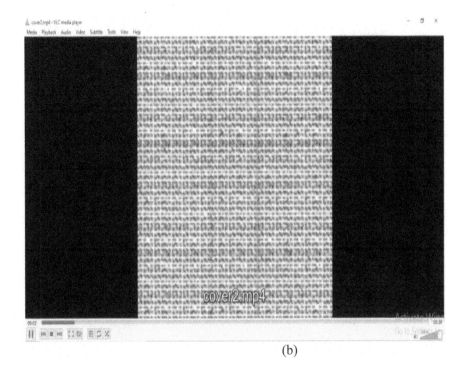

(b)

FIGURE 6.10 (Continued)

6.5.7 CLUSTERED FRAMES IN COVERED VIDEO

Video name: Container_qcif

Video name: Forman_qcif

The statistical results obtained to show the performance of the proposed secure communication model with DBSCAN clustering in contrast to existing k-means clustering model were shown in Table 6.1. The resultant performance metrics are shown in Tables 6.2 and 6.3.

TABLE 6.2

Relative PSNR Comparison after Encryption

Video Sequence (cover)	Relative PSNR				
	Before Encryption (db)	After Encryption (db)	Proposed	Existing (17)	Existing (2)
Container_qcif.yuv	40.30	37.44	1.0763	1.0189	1.0309
Foreman_qcif.yuv	39.70	37.69	1.0533	1.0306	1.0338
Carphone_qcif.yuv	38.73	36.92	1.0490	1.0007	1.0237
Stefan_qcif.yuv	37.86	36.63	1.0335	1.0120	1.0101

TABLE 6.3
NPCR and UACI Values for the Proposed Model

Video Sequence (cover)	NPCR (%)	UACI (%)
Container_qcif.yuv	99.14	33.42
Foreman_qcif.yuv	99.54	41.31
Carphone_qcif.yuv	99.49	38.92
Stefan_qcif.yuv	99.47	38.21

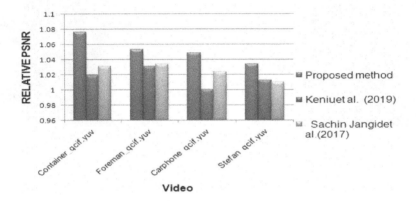

FIGURE 6.11   Relative PSNR after Encryption

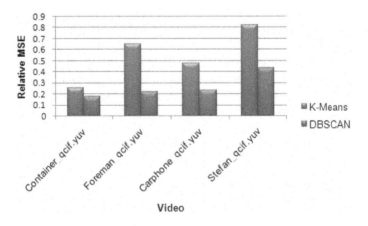

FIGURE 6.12   MSE Comparison.

The relative PSNR obtained after the encryption process in comparison with the existing models is shown in Figure 6.11.

The performance of the proposed model with DBSCAN clustering and existing clustering approach those relative values are shown in Figures 6.12, 6.13 and 6.14.

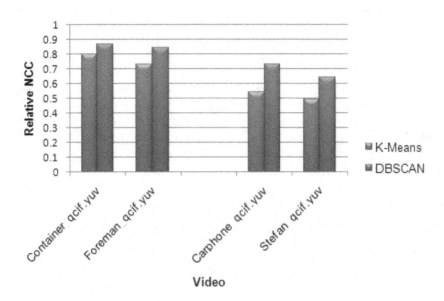

FIGURE 6.13   NCC Comparison.

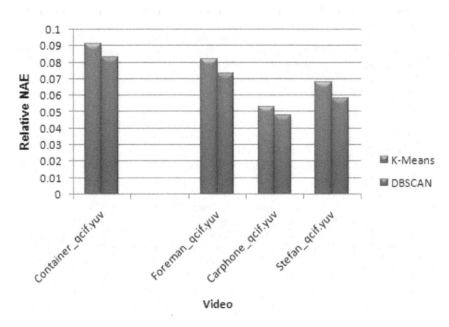

FIGURE 6.14   NAE Comparison.

6.6 CONCLUSION

In contrast to cryptography, steganography uses videos or images for concealing the sensitive data. As a result, steganography enforces more security than do cryptography techniques, which are prone to cryptanalysis. With video steganography, we can reduce the risk of attacks on the data or information to be hidden. The Arnold cat mapping technique is used for creating chaos in a stego-video, which further increases the security. Although the distortion of the frames is a main issue, it contributes to the security of stego-video. This model achieves a better performance in comparison to existing approaches and adds some extra security in wireless data transmission using deep learning.

REFERENCES

1. Abomhara, M., O. Zakaria, O. Khalifa. (2010). An Overview of Video Encryption Techniques. *International Journal of Computer Theory and Engineering*, 2, 103–110.
2. Jangid, S., and S. Sharma. (2017). High PSNR Based Video Steganography by MLC (Multi-Level Clustering) Algorithm. *Proceedings of ICICCS*, 589–594, June.
3. Li, J., X. Yang, and X. Liao. (2016). A Game-Theoretic Method for Designing Distortion Function in Spatial Steganography. *Multimedia Tools and Applications*, 76(10), 12417–12431.
4. Provos, N., and P. Honeyman. (2017). Hide and Seek: An Introduction to Steganography. *IEEE Security Privacy*, 1.
5. Zhang, Y., D. Ye, J. Gan, Z. Li, and Q. Cheng. (2018). An Image Steganography Algorithm Based on Quantization Index Modulation Resisting Scaling Attacks and Statistical Detection. *Computers, Materials & Continua*, 56(1), 151–167.
6. Jie, Wang, X. Jia, X. Kang, &Y. Shi. (2019). A Cover Selection HEVC Video Steganography Based on Intra Prediction Mode. *IEEE Access*, 7, 119393–119402.
7. Ramathan, J. et al. (2017). A Robust and Secure Video Steganography Method in DWT-DCT Domains Based on Multiple Object Tracking and ECC. *IEEE Access*, 5, 5354–5365.
8. Zhang, Y., M. Zhang, X. Yang, D. Guo, and L. Liu. (2017). Novel Video Steganography Algorithm Based on Secret Sharing and Error-Correcting Code for H.264/AVC, *Tsinghua Science and Technology*, 22(2), 198–209.
9. Zhang, Zuho, G. Fu, R. Ni, J. Liu, and X. Yang. (2020). A Generative Method for Steganography by Cover Synthesis with Auxiliary Semantics. *Tsinghua Science and Technology*, 25(4), 516–527.
10. Emad, Elshazly, A. Safey, A. Refaat, Z. Osama, E. Sayed, and E. Mohamed. (2018). A Secure Image Steganography Algorithm Based on Least Significant Bit and Integer Wavelet Transform. *Journal of Systems Engineering and Electronics*, 29(3), 639–649.
11. Hu, Donghui, L. Wang, W. Jiang, S. Zheng, and B. Li (2018). A Novel Image Steganography Method via Deep Convolutional Generative Adversarial Networks. *IEEE Access*, 6, 38303–38314.
12. Kaixi, Wang, and Q. Gao. (2019). A Coverless Plain Text Steganography Based on Character Features. *IEEE Access*, 7, 95665–95676.
13. Ayesha Saeed, Fawad, and M.J. Khan. (2020). An Accurate Texture Complexity Estimation for Quality-Enhanced and Secure Image Steganography. *IEEE Access*, 8, 21613–21630.
14. Xinghong, Li B., S. Tan, and J. Zeng. (2019). A Novel Steganography for Spatial Color Images Based on Pixel Vector Cost. *IEEE Access*, 7, 8834–8846.

15. Ming Guo, Jing, D. Riyono, and H. Prasetyo. (2018). Improved Beta Chaotic Image Encryption for Multiple Secret Sharing. *IEEE Access*, 6, 46297–46321.

16. Li, Shan-Shani. (2020). An Improved DBSCAN Algorithm Based on the Neighbor Similarity and Fast Nearest Neighbor Query. *IEEE Access*, 8, 47468–47476.

17. Ke Niu, Jun et al. (2019). Hybrid Adaptive Video Steganography Scheme Under Game Model. *IEEE*, 7, 61523–61533.

18. Missile Launch Video. (2007). *South Korea Missile Test* [video file], 28 August, www.youtube.com/watch?v=6IWpJUb5CTs.

19. *Test Dataset for Container Videos* [video file], https://media.xiph.org/video/derf/y4m/.

7 Target Prophecy in an Underwater Environment Using a KNN Algorithm

Menaka D, Sabitha Gauni, C.T. Manimegalai, K. Kalimuthu, and Bhuvan Unhelkar

CONTENTS

7.1 INTRODUCTION

Wireless sensor networks (WSNs) typically comprise several sensor nodes with small size, low power consumption and low resources. The nodes scattered around the target sensor node efficiently collect data from their environment and forward the sensed data to control structures, also known as sinks or base stations. These sensor nodes are fitted with a range of sensors, such as acoustics, thermal, chemical, heat, optical, and temperature and have a tremendous capability to develop effective applications. WSNs in underwater are an upcoming technology used in

DOI: 10.1201/9781003107477-7

finding lost debris or treasures in deep water. The location of the source in an underwater transducer provides a successful path for extending ocean acoustics. Numerous attempts have been made to solve the problem of underwater localization with more accuracy. This is a challenge for designing appropriate algorithms for underwater applications. Specific problems that researchers need to solve include data stability, data aggregation, clustering of nodes, venue, scheduling of activities, energy-conscious routing, protection and identification of faults, lost debris, or treasure. Machine learning (ML) was introduced as an artificial intelligence (AI) technique during the late 1950s (1). This technique evolved very rapidly, and its algorithm renders it computationally feasible and efficient for identifying deep-water locations. ML approaches were commonly utilized in bioinformatics, speech recognition, spam analysis, computer vision, fraud detection, and advertisement networks. These ML tasks include classification, regression, and intensity estimation. Because ML can give the optimized result in locating the sensor nodes, their algorithms for deep-water sensing are based on the following techniques:

1. It builds computational models for active learning and solves the question of acquiring information while enhancing the efficiency of existing systems (2).
2. It establishes analytical methods for identifying and explaining consistencies and trends in training data to enhance computer performance (3).

From the preceding techniques, the method of calculating the nodes and elements of a network's spatial coordinates named localization can be achieved. This is a significant capability of a sensor network, as most sensor network operations involve node positioning (4). Many large-scale networks utilize Global Positioning System (GPS) equipment to localize themselves in each node. That's costly because GPS needs electricity. Because electromagnetic signals do not propagate deep in water the FPS performance suffers in underwater systems. The k-nearest neighbor (KNN) algorithm is preferred for the simplest way of predicting the targeted sensor nodes in underwater. KNN is an optimization technique used in maximizing output, identifying patterns, processing images, etc. For algorithms with broad training data set (5), the KNN method is suitable. The KNN algorithm has reliable outcomes of optimization and seeks to identify new creations based on characteristics and examples in training (6). KNN can be adjusted to calculate the percentage of data used for evaluation, the number of a k neighbor, and classification of the data were collected before use. The data are standardized and simplified before grouping to increase the accuracy of KNN in diagnosis and prediction. This chapter aims at predicting the targeted sensor node underwater using the KNN algorithm for localization. KNN is a random method with high accuracy and easy implementation. KNN is not used in underwater environments because of its nonlinearity and complexity in executing a large data set. Here, with the help of its neighboring sensor nodes, the location of the targeted sensor node underwater is determined by using a KNN algorithm. The result is based on the accuracy of finding the object and the error that occurred during the localization.

7.2 LOCALIZATION ALGORITHMS FOR INDOOR APPLICATIONS

Several fundamental localization methods are outlined as follows for WSNs utilizing ML.

7.2.1 BAYESIAN NODE LOCALIZATION

Morelande et al. (7) have been focusing on designing a sensor network localization scheme using anchor nodes. It works on improving the methodology of progressive correction to estimate samples from the likelihood of becoming closer to the subsequent probabilities. This method is easily applicable to the localization of broad networks, that is, consisting of several thousand sensor nodes. The revolutionary idea of using the Bayesian identify method is promising because by analyzing prior information and chances, it can manage unfinished data sets.

7.2.2 ROBUST POSITION IDENTIFICATION OF OPERATION

Lu and Fu (8) aimed to fix the sensor issue and the localization of operation in smart homes. Such fascinating hobbies include listening to the radio, utilizing smartphone or computer packs, opening or closing the refrigerator, or learning, among others. The developers need to comply with both the environmental and human constraints in these sensor applications to maintain their ease of service. This proposed system is referred to as Ambient Intelligence Compatible Interface Alternative, which enables human contact with electrical devices used in our homes in an effective manner, such as automated control of power supply. This approach is a flexible localization tool, but it relies on implementations, so the developers will also have a range of approved activities. User learning apps are picked and tested manually according to the tasks and the area of interest. Unmonitored predictive analytics algorithms, such as deep learning methods and the non-convex factorization algorithm, are suggested for the automated extraction of features, which is to (9) be investigated in the centralized system to solve the previously discussed problem.

7.2.3 NEURAL NETWORK LOCALIZATION

Shareef et al. (10) essentially contrast and address the localization approaches dependent on various forms of neural networks. Centered on Neural Network Localization, this analysis compares location using multilayer perceptron (MLP), recurrent neural networks (RNNs), and radial base function (RBF) and indicates that the RBF neural network performs with low place error at the cost of high available resources. On the other hand, the MLP performs with low computing and storage power usage. Similarly, two types of anchor node localization algorithms utilizing a received signal strength indicator (RSSI) were presented by Yun et al. (11). The first class uses a fuzzy logic method and genetic algorithm while the second type uses a neural network as input nodes to determine the position of the sensor node using RSSI observations from all anchor nodes. Similarly, Chagas et al. (11) used neural networks as a learning network reference for RSSI localization. The key benefit of these neural network–dependent localization algorithms lies in their ability to provide location in the form of continuously valued vectors.

7.2.4 LOCALIZATION IN SUPPORT VECTOR MACHINE (SVM)

For the localization of sensor nodes, the SVM approach has been largely facilitated where it is unrealistic to connect a peer-positioning system to each sensor node by implementing SVM and networking information technologies, Yang et al. (12) suggested a localization strategy for mobile nodes in this way. The suggested method utilizes the radio-frequency oscillation, such as the RSSI measure, in its initial step to explore variations in node movement. SVM will be performed for motion detection to include the current spot. Tran and Nguyen have focused on localization strategies based on vector machines help identify nodes in sensor applications.

7.2.5 LOCALIZATION USING A SELF-ORGANIZING MAP

Paladina et al. (13) proposed WSN positioning solutions which consist of thousands of nodes using the self-organizing map (SOM). The proposed algorithm is implemented in each node using a basic SOM algorithm, consisting of a 3 × 3 input layer attached to the two output-layer neurons. Specifically, the input layer is constructed using the position information of eight anchor nodes corresponding to the unknown node. The output layer is used after adequate training to display the spatial coordinates of the unknown node in a two-dimensional region. The basic drawback of this approach is that node installation will be uniformly distributed over the region being tracked.

7.2.6 PASSIVE UNDERWATER TARGET TRACKING

The implementation of the conditionally minimax nonlinear filtering (CMNF) algorithm to the web-based prediction of aquatic vessel motion was demonstrated by Andrey Borisov et al. (14) with the integration of sound navigation and ranging (sonar) and Doppler discrete-time noisy sensor observations. There are the following different benefits to the projected CMNF. First, the created figures are unbiased. Second, the logical covariance matrix of the CMNF mistakes suits the real values. Third, to significantly increase the precision of the measurement, the CMNF algorithm provides the possibility of choosing the conversion of recently experienced, basic estimation, and corrective functions in any specific case of the monitoring process. The method provides the results of the CMNF algorithm applied to the solution of the evasive maneuvering autonomous underwater vehicle (AUV) tracking problem by integrating the bearing-Doppler measurements.

7.2.7 PREDICTION OF UNDERWATER TARGET USING SONAR

Harvinder Singh et al. (15) explore the sonar technique, which relays the barrier or the surface is a rock or a mine on certain criteria to be capable of detecting. This criterion is expressed through the developments in predictive analytics. The main objective is to emit a capable representative of prediction, united by algorithmic characteristics of machine learning, which can decide whether either a rock or a mine or some other organism or some other form of body is the subject of the sound wave. This attempt is a clear-cut case study that comes up with a rock and mineral grading machine learning plan, performed on a massive, highly spatial and complex sonar data set. The attempts are carried out on a highly spatial sonar data set and an accuracy of 83.17 percent has been achieved and AUC has been 0.92 with the least error in order to further elaborate this prediction model.

7.2.8 Target Tracking in Underwater Acoustic Sensor Network

Divin Ganpathi et al. (16) discuss how, with the depletion of earthly resources, human beings are moving to explore marine resources. In a marine climate, various kinds of difficulties are encountered. Acoustic waves are used in underwater civilian and military applications for communication devices. Submarine acoustic target monitoring is an important part of marine exploration. Based on the nature of the marine world, a large number of target monitoring techniques are being used. This aims to explore a survey of recent studies on technologies for underwater monitoring. The classification of different underwater target tracking techniques is performed depending on the methods used. The algorithms are tested to find the most effective one for underwater target recognition. Some potential problems of target monitoring with multiple objectives are tracking, measurement accuracy enhancement, data fusion, etc.

7.2.9 Prediction and Estimation of Underwater Target Using Multichannel Sonar

A new approach is proposed by Nick H. Klausner et al. (17) to predict and estimate the performance with multiple synthetic aperture sonar for a multichannel detection system. The prediction and estimation of the output are achieved by analyzing the statistics for multichannel coherence and characterizing the context conditions within the image. For different context circumstances, it tests the system's ability to provide an assessment of image sensitivity. The null distribution of statistical test is used to estimate the experimental saddle point estimation to select the cutoff and to achieve the required rate of false alarm. Test data are summarized on two real and one synthetic sonar visualization data sets with different focus and environment conditions, showing the ability of the proposed methods to explain the distribution of the likelihood ratio and predict the detector performance in low- to medium-clutter conditions.

The preceding localization algorithms give the various applications and algorithms used in the indoor environment, underwater applications using sonar, self-organizing maps and localizations in neural networks.

7.3 ML

ML is an analysis of computational algorithms or programs that eventually develop through training or experience. ML is a representation of artificial intelligence (AI) to make assumptions or decisions. ML algorithms are developed based on the training data set, without being explicitly programmed. The aim of ML is to make the computers learn and behave like humans from the provided training data set. ML is broadly classified into three categories:

a. The "supervised ML algorithms" is an approach learned from the past to predict future values. The algorithm generates an approximate function from the recognized training data set to estimate the expected output. The outcoming results can be compared with the expected output to identify the error and rectify it according to the model.
b. The "unsupervised ML algorithm" is a contrast model compared to the supervised learning algorithm. This model is used when the information provided is neither labelled nor classified. The unsupervised approach describes how to discover a hidden structure from the unlabeled data.
c. The "reinforcement ML algorithm" is a method of learning that communicates with the world by creating behavior patterns and discovering mistakes or benefits. The most important characteristics of reinforcement learning are trial and error search and back propagation. To optimize its efficiency, this approach enables machines and programming environments to automatically evaluate the optimal behavior within a limited instance. For the agent to learn which behavior is better, simple reward feedback is required; this is known as the reinforcement signal.

7.4 KNN NODE PREDICTION

KNN is an optimization technique in different fields, such as maximizing output, identifying patterns, processing images, and so on. For algorithms with broad training data, the KNN solution is suitable. The KNN algorithm has correct outcomes of optimization and seeks to identify a new classification based on characteristics and samples in training (18, 19). In the KNN process, data are divided into two sets (i.e) training set and testing set. The KNN classifier which is calculated based on the distance determines the validity of the process set. The KNN method is often seen in forecasts of stock prices. The comparison data used on the Jordanian stock exchange are six major enterprises. The findings of this analysis can be used to help provide investors with appropriate investment information. One of the slow analysis tools most commonly used is the KNN algorithm (19).

It is an efficient algorithm for differentiation that can cope with complex problems. If there are hundreds of features and the preparation set is large in size, the use of the classification model can well classify the samples. The key concept of KNN is to identify a sample data x, to compute the distance of all x, and to obtain the k closest neighbors in the training data analysis observations. The class label representing the maximum number of nearest neighbors to reference x is referred to by KNN. In terms of Euclidean distance, the closest neighbor is given as

$$d\left(y_{i,},y_{j}\right)=\left(\sum_{s=1}^{m}\left|y_{is}-y_{js}\right|^{2}\right)^{\frac{1}{2}},\qquad(7.1)$$

where y_i and y_j represents the vector space of two samples indicating the attributes s as y_{is} and y_{js}, respectively. The KNN process is assumed to be among the easiest classification algorithms. The KNN algorithm is a component of the data set into two called the training set and the test package. Consider the training set $\left[\left(x_i, y_j \right)_{i=1}^{N} \right] \equiv D$ [$(x_i,y_i)_{i=1}^N$]; D with x_j is the training vector with n-dimensional set and y_j is the respective class label.

a. The Euclidean distance d between the X' and each of (x_j, y_j) is calculated:

$$d\left(x, x_i\right) = x - x_{jL2}.$$ (7.2)

b. The calculated distances are sorted in ascending order.
c. The majority of the vote of the nearest neighbor is predicted based on x' class label:

$$y' = argmax_x \sum \left[x_j, y_j \right] \equiv D'W_i\delta\left(x \equiv y_j'\right).$$ (7.3)

Next, the properties and associated aspects of the samples in the training study are described in an m-dimensional format. This implies that all the measurements are converted into the m-dimensions vector space. The test-set images, however, frequently scatter in the m-dimensional vector space but with undefined features. Once a test sample is given, the KNN classifier will search the k preparation experiments put alongside the unknown test sample. The research data are then allocated to the type that has the largest number in the corresponding k (20, 21). As a commonly used algorithm in machine learning, KNN has not only several advantages and optimistic accuracy performance but also two disadvantages. Hundreds of instances, tests in sparsely or densely distributed vector space that cause these various k-values, may lead to various results. As a result, it is difficult to decide which k is sufficient for classification. Measure the closest research sample neighbor's k. It may not be efficient as it may be merely dependent on Euclidean distance. In light of the preceding questions, we need to create potential strategies for coping with multidimensional problems and conditions of unequal distribution. And our men this article focuses on the application of the kernel process and Reduce the attribute to enhance precision.

7.5 UNDERWATER LOCALIZATION USING KNN

KNN is a supervised machine learning algorithm focused on the classification of data samples. This algorithm measures the average measurements of adjacent sensor nodes and estimates the missing sensor values. KNN uses the Euclidean distance equations for measuring the relative point between various sensors. Hence, high computing capacity (22, 23) is not needed. Throughout the area of localization and entity tracking, incident identification and demand handling, network access management, protection and intrusion detection and service of quality, data confidentiality, and fault detection (24, 25). The KNN can solve problems in classification and regression methods. The classification approach involves the prediction of a target (26, 27) with the given unlabeled data. The regression approach is to calculate the average of the numerical targets of the k neighbor.

Underwater localization in KNN algorithm comes under the regression approach. The working of the KNN algorithm underwater is explained in Figure 7.1. To implement any algorithm in ML, first, the data set has to be uploaded. Therefore, as given in Figure 7.1, the k value is initialized; that is, the nearest node as the integer is chosen and the data set is trained according to the requirement. This is done under the training part. In the testing part, the distance between the nodes is calculated using the Euclidean distance. Now the calculated distance is sorted according to the ascending order. Once the order is sorted, the k value is assigned to find the target node position, that is, the node neighbor to the targeted node. With the help of this process, the nearest nodes of the target are identified, and the information is gathered from the predicted node. This approach helps with finding the target location using the KNN algorithm. The applications underwater include the target estimation and missing debris in ocean climate.

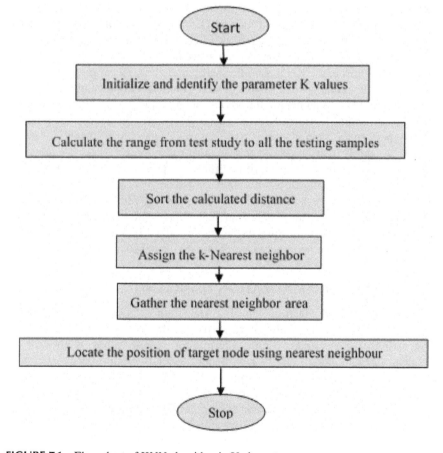

FIGURE 7.1 Flow chart of KNN algorithm in Underwater

TABLE 7.1

Underwater Target Node Prediction Using the KNN Algorithm

Target Node	Estimated Nearest Neighbor Node
12	15
8	4
4	8
14	13
5	10
11	14

7.6 RESULTS AND DISCUSSION

In underwater wireless sensor networks, two issues are addressed in this chapter:

1. When the deployment for sensor nodes is done in underwater, the nodes cannot be observed frequently by visiting the environment. In such case, the prediction of alive and dead nodes, that is, active and inactive node prediction, underwater is necessary to collect the information.
2. After identifying the live status of nodes (i.e., active or not), the next step is to predict the position of the target sensor present in underwater. This includes determining the path for localization using an AUV and sonar.

7.6.1 ACTIVE AND INACTIVE NODE PREDICTION

The active and inactive node prediction using KNN algorithm is simulated with a data set containing one month of underwater temperature data. As first step, the number of sensors that are deployed underwater is plotted in Figure 7.2.

Next, to predict active and dead node, a data set having a combination of active and inactive (dead) nodes is used. As explained in Figure 7.1, the data set is trained and tested in the ratio of 20:80 by the KNN prediction technique. The distance and the location of the nodes is calculated using the Euclidean method and the outcomes are noted. By this method, the outcome 1 is predicted to be the active node, and the result with an outcome 0 is predicted to be the inactive (dead) nodes as plotted in Figure 7.3. The active nodes are the nodes that are alive from the time of deployment and the information providers till date. The inactive nodes are the dead nodes that do not provide the information. The nodes may become inactive due to a lack of energy in the sensor, heavy water current, underwater mammals or marine activities like shipping or theft. From the simulated result, the training accuracy is obtained to be 96.333%, that is, correctly predicted observation, and 98.132% f1 score, that is, testing accuracy as shown in Table 7.2.

7.6.2 TARGET LOCATION WITH ITS NEIGHBOR NODES

The position of the target node is calculated with the nearest neighbor using the KNN algorithm. Underwater, as shown in Figure 7.2, 50 static nodes are deployed in

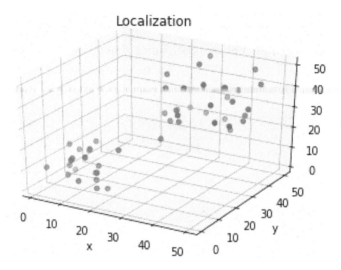

FIGURE 7.2   Deployment of Sensor Nodes in Underwater.

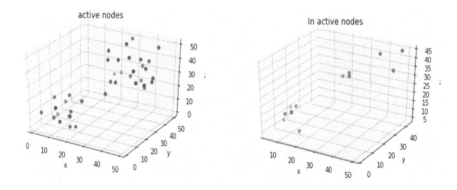

FIGURE 7.3   Prediction of Active and Inactive Nodes Using KNN.

TABLE 7.2
Qualitative Analysis Using the KNN Algorithm

Parameters	Qualitative Analysis Using KNN
Accuracy (training set)	0.9633
Recall score	1.0
f1 score (testing set)	0.98

---

a three-dimensional axis at a 10-mts depth, and their estimated locations are found using the KNN algorithm. The distance between each node is calculated using the Euclidean distance technique. The neighboring node is determined, based on the Euclidean results, as seen in Table 7.1. Here, the temperature data set is trained to initialize the k value and tested to estimate the location of the targeted sensor node with the training accuracy as 0.9633; recall the score, that is, sensitivity, is 1.0 and the testing accuracy, that is, f1 score, is 0.98, as shown in Table 7.2. As the k value to locate the neighbor node is raised, this contributes to an improvement in the MSE in the center. This error is determined to evaluate the disparity between the values expected and those observed. The root mean square error (RMSE) is plotted in Figure 7.4. In other terms, it is denoted as the quality of fit between the actual and predicted models. From Figure 7.4, the value of the RMSE is high when the k value is less, the error is low when the k value is moderate, and it gradually increases when the k value is high. Depending on the amount of information in the data set, the k value is estimated. If the data set has a huge number of samples, the k value can be increased for high accuracy.

Lower RMSE values indicate that the model is a better fit at k = 8 to k = 15. This is a good measure of how the model is accurately predicted, which is an important criterion for target prediction or estimation. So, from the results, the KNN algorithm is a better fit for underwater conditions. Even though many fluctuations are seen in the data set, the KNN algorithm predicts the target location with its neighbor nodes in underwater conditions with a better accuracy of about 96.33% and a lower computational speed of less than a few milliseconds. From the obtained results, the targeted sensor node prediction underwater can be effectively localized by an AUV or sonar using the KNN algorithm to estimate the path to locate the sensor.

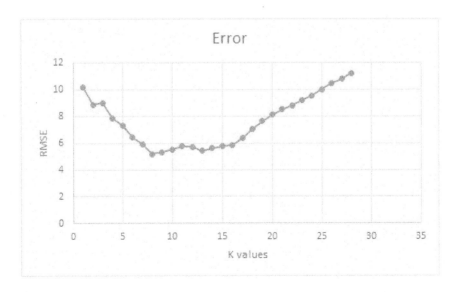

FIGURE 7.4   The Curve Plotted for the k Value versus the RMSE in Localization.

7.7 CONCLUSION

From the preceding simulation, the output of the localization is directly linked to the data available for ML training. ML is a promising tool and can be more efficient and successful than other algorithms if adequate training data are provided. In the case of the absence of an adequate training data set, ML might not be a feasible choice. In this chapter, a novel underwater application using the KNN algorithm was proposed that aims at pairing ML with an acoustic propagation approach, since the underwater application is not tested using KNN because of its large data set. This chapter focused on the computation of the KNN using a large data set with less computational time. This improves the prediction of the targeted sensor nodes underwater to estimate the sensor node position using the nearest neighbor. From these results, KNN can classify whether the nodes deployed months or years before are active or inactive at present to analyze its accuracy and error of prediction. This emphasizes the fact that inactive nodes can be replaced for a high rate of prediction in lost debris. The error rate was proved to be less in the case of the parametric values of k = 8 to k = 15. Experimental results using KNN had a proved 96.33% accuracy when it was used for underwater target prediction for locating the position of sensor node.

REFERENCES

1. Duffy, A. H. B. (1997). The 'What' and 'How' of Learning in Design. *IEEE Expert*, 12(3), 71–76, May. doi:10.1109/64.590079.
2. Liu, Hui, Houshang Darabi, Pat Banerjee, and Jing Liu. (2007). Survey of Wireless Indoor Positioning Techniques and Systems. *IEEE Transactions on Systems, Man and Cybernetics, Part C (Applications and Reviews)*, 37(6), 1067–1080, November. doi:10.1109/tsmcc.2007.905750.
3. Jayaraman, Prem Prakash. (2010). *Trends in Applied Intelligent Systems*. Berlin, Heidelberg: Springer.
4. Lee, Tae Choong, and Sang Hak Chung. (2005). *Artificial Intelligence and Simulation*. Berlin, Heidelberg: Springer.
5. Cayirci, Erdal, Hakan Tezcan, Yasar Dogan, and Vedat Coskun. (2006). Wireless Sensor Networks for Underwater Survelliance Systems. *Ad Hoc Networks*, 4(4), 431–446, July, doi:10.1016/j.adhoc.2004.10.008.
6. Lee, Sang Hak, and Tae Choong Chung. n.d. *Data Aggregation for Wireless Sensor Networks Using Self-Organizing Map*. Berlin, Heidelberg: Springer. Accessed 2005.
7. Morelande, M. R., M. Brazil, and B. Moran. (2008). Bayesian Node Localisation in Wireless Sensor Networks. *2008 IEEE International Conference on Acoustics, Speech and Signal Processing*, 2545–2548.
8. Doucet, Arnaud, Nando De Freitas, and Neil Gordon. (2001). *Sequential Monte Carlo Methods in Practice*. New York: Springer.
9. Wang, Yu-Xiong, and Yu-Jin Zhang. (2013). Nonnegative Matrix Factorization: A Comprehensive Review. *IEEE Transactions on Knowledge and Data Engineering*, 25(6), 1336–1353, June, doi:10.1109/tkde.2012.51.
10. Yun, Sukhyun, Jaehun Lee, Wooyong Chung, Euntai Kim, and Soohan Kim. (2009). A Soft Computing Approach to Localization in Wireless Sensor Networks. *Expert Systems with Applications*, 36(4), 7552–7561, May, doi:10.1016/j.eswa.2008.09.064.
11. Chagas, S. H., L. L. de Oliveira, and J. B. Martins. (2012). An Approach to Localization Scheme of Wireless Sensor Networks Based on Artificial Neural Networks and Genetic Algorithms. *10th IEEE International NEWCAS Conference*, 137–140.

12. Yang, Bin et al. (2007). Area Localization Algorithm for Mobile Nodes in Wireless Sensor Networks Based on Support Vector Machines. In *Mobile Ad-Hoc and Sensor Networks*. Berlin, Heidelberg: Springer, 561–571.

13. Peng, Mugen, Zhiguo Ding, Yiqing Zhou, and Yonghui Li. (2012). Advanced Self-Organizing Technologies Over Distributed Wireless Networks. *International Journal of Distributed Sensor Networks*, 8(12), 821982, 24 December, doi:10.1155/2012/821982.

14. Borisov, Andrey et al. (2020). Passive Underwater Target Tracking: Conditionally Minimax Nonlinear Filtering with Bearing-Doppler Observations. *Sensors*, 20(8), 2257, 16 April, doi:10.3390/s20082257. Accessed 3 November 2020.

15. Singh, H., and N. Hooda. (2020). Prediction of Underwater Surface Target Through Sonar: A Case Study of Machine Learning. In A. Chaudhary, C. Choudhary, M. K. Gupta, C. Lal, and T. Badal. (eds), *Microservices in Big Data Analytics*. Singapore: Springer, 111–117.

16. Ganpathi, T. Divin. (2020). Review on Target Tracking Methods for Underwater Acoustic Sensors. *Journal of Mechanics Of Continua and Mathematical Sciences*, 15(2), 27 February, doi:10.26782/jmcms.2020.02.00031. Accessed 3 November 2020.

17. Klausner, Nick H., and Mahmood R. Azimi-Sadjadi. (2020). Performance Prediction and Estimation for Underwater Target Detection Using Multichannel Sonar. *IEEE Journal of Oceanic Engineering*, 45(2), 534–546, April, doi:10.1109/joe.2018.2881527. Accessed 3 November 2020.

18. Zhang, Xueying, and Qinbao Song. (2014). Predicting the Number of Nearest Neighbors for the K-NN Classification Algorithm. *Intelligent Data Analysis*, 18(3), 449–464, 30 April, doi:10.3233/ida-140650.

19. Xing, Wenchao, and Yilin Bei. (2020). Medical Health Big Data Classification Based on KNN Classification Algorithm. *IEEE Access*, 8, 28808–28819, doi:10.1109/access.2019.2955754.

20. Li, Yong-hua. (2008). An Improved Algorithm for Attribute Reduction Based on Rough Sets. *Journal of Computer Applications*, 28(8), 2000–2002, 20 August, doi:10.3724/sp.j.1087.2008.02000.

21. Datta, Amrita, and Mou Dasgupta. (2020). On Accurate Localization of Sensor Nodes in Underwater Sensor Networks: A Doppler Shift and Modified Genetic Algorithm Based Localization Technique. *Evolutionary Intelligence*, 7 January, doi:10.1007/s12065-019-00343-1. Accessed 29 April 2020.

22. Ganpathi, T. Divin. (2020). Review on Target Tracking Methods for Underwater Acoustic Sensors. *Journal of Mechanics of Continua and Mathematical Sciences*, 15(2), 27 February, doi:10.26782/jmcms.2020.02.00031. Accessed 3 November 2020.

23. Guo, Ruolin et al. (2020). Mobile Target Localization Based on Iterative Tracing for Underwater Wireless Sensor Networks. *International Journal of Distributed Sensor Networks*, 16(7), July, doi:10.1177/1550147720940634. Accessed 4 November 2020.

24. Rauchenstein, Lynn T. et al. (2018). Improving Underwater Localization Accuracy with Machine Learning. *Review of Scientific Instruments*, 89(7), 074902, July, doi:10.1063/1.5012687. Accessed 4 November 2020.

25. Su, Xin et al. (2020). A Review of Underwater Localization Techniques, Algorithms, and Challenges. *Journal of Sensors*, 1–24, 13 January, doi:10.1155/2020/6403161. Accessed 4 November 2020.

26. Tsai, Pei-Hsuan et al. (2017). Hybrid Localization Approach for Underwater Sensor Networks. *Journal of Sensors*, 1–13, doi:10.1155/2017/5768651. Accessed 22 May 2020.

27. Nair, Saranya, and Suganthi K. (2020). Energy Efficient 4 Dimensional Heterogeneous Communication Architecture for Underwater Acoustic Wireless Sensor Networks. *International Journal of Scientific & Technology Research*, 9(1), January.

8 A Model for Evaluating Trustworthiness Using Behaviour and Recommendation in Cloud Computing Integrated with Wireless Sensor Networks

Lokesh B. Bhajantri and Tabassum N. Mujawar

CONTENTS

DOI: 10.1201/9781003107477-8

8.1 INTRODUCTION

Cloud computing introduces the term *computing as a service*. Cloud computing provides different computing resources to the users as a service over the internet. In recent years, many organizations are also rapidly moving towards cloud computing. The major reason behind this is that both users and organizations benefit in many ways. The primary benefits include a reduction in operation cost, less capital investment, more scalability, much less maintenance, endless capability and others. There are abundant computing resources available for users as a shared pool, from which users can select as per their demand. The key benefit for the user is that it is a pay-per-usage model. Hence, users will pay only for their consumption of service instead of investing a large cost for purchasing the resources. The services are fully managed by the cloud service provider (CSP). The delivery of required services and resources to all cloud users is handled by the service provider.

Wireless sensor networks (WSNs) are also a promising research field that has interconnected sensor nodes, which collect data about the surrounding environment. These data can be processed and utilized for different applications. The sensor nodes have limited memory and power capability. If a large amount of data is extracted from sensor nodes, then there must be a proper mechanism to store these data securely so that the collected data can be used in the future to retrieve additional knowledge, to be utilized for any particular application. The various issues that have an impact on the performance of WSNs include communication characteristics, medium access schemes, localization, synchronization, data aggregation and data dissemination, database-centric querying, quality of service (QOS)–based communication and computing, security, dealing with denial-of-service attacks, security in routing and node capture issues, among others (1). In this chapter, an approach to integrate the cloud environment and WSNs is proposed for storing, processing and accessing the sensor data efficiently and securely in a cloud environment.

With the help of cloud computing, users can access a variety of resources from anywhere and at any time with ease and less management. Compared to the traditional system, cloud computing offers many features and benefits. But many users and organizations are still not adopting cloud computing and consider the security concerns as main reason for why. The data stored in the cloud are exposed to other users in the cloud environment, and users lose control over the data. Users need to trust the service providers as they are responsible for keeping the data secure in the cloud environment. Although many research works are presented to address the security challenges present in a cloud environment, there are still some open issues. The main issue is that users are not ready to fully trust the services offered by cloud computing; hence, there is a limitation on the wide adoption of cloud computing services. Therefore, it is necessary to build a trust relationship between users and the cloud environment so that users will be guaranteed to get trustworthy service from the cloud. On the other hand, it is equally important to verify that the user accessing the cloud service or resource is a legitimate user. To provide proper and secure access to cloud resources, the trust mechanism must be incorporated into the traditional method. Users will trust any cloud service only if the performance of that particular service will satisfy users' needs. The performance can be monitored in terms of different QoS attributes. The relative importance of each QoS attribute should be taken into consideration while assigning trust value to any cloud service. Also, the feedback given by users after interacting with a cloud service will also play a major role in the assessment of the trustworthiness of the cloud service. In this chapter, we address the issue of efficient assessment of trust levels of a cloud service and cloud users both by using a machine learning algorithm. An algorithm to quantify trust levels for cloud services and cloud users based on user preferences, QoS attributes, service-level-agreement (SLA) parameters, recommendations and behaviour is presented in the chapter. The proposed method applies the fuzzy analytical hierarchy process (AHP) method to assign weights for different attributes considered in the trust calculation. Depending on the user preference, the method decides the relative importance among the various attributes. The paper also implements fuzzy c-means clustering algorithm to group users and cloud services into different clusters and assigns them the appropriate trust levels. The performance of the proposed method is compared with existing trust models in terms of accurate computation of trust levels.

The major contributions of this chapter are as follows:

1. The method to prioritize among the different performance attributes, SLA parameters and user behavior attributes by applying a fuzzy AHP algorithm is proposed.
2. The cloud service trust computation method based on performance, SLA and recommendation (CSTM_PSR) that applies fuzzy c-means clustering algorithm to divide cloud services into different groups is presented.
3. The cloud user trust computation method based on behavior and recommendation (CUTM_BR) to group cloud users into different clusters by applying a fuzzy c-means clustering algorithm is presented.

The rest of the chapter is organized as follows: Section 8.2 presents the existing trust evaluation techniques for users and services in a cloud computing environment. Section 8.3 gives complete insights into the proposed method of trust computation for cloud services and cloud users in a cloud environment. Section 8.4 discusses the simulation and results of the proposed methods. Finally, conclusion is presented in Section 8.5.

8.2 RELATED WORK

The trust-based model using machine learning for authorizing users is presented in Khilar et al. (2). In this model, the users are assigned priorities based on the computed trust value and then the resources are allocated to them. The paper also presents a method for evaluating the trustworthiness of cloud resources. The different machine learning algorithms are implemented to compute trust levels for users and resources. The trust-based cloud service selection model is proposed in Yubiao Wang et al. (3). The hierarchical clustering method is used to categorize user preferences and evaluate the trust value for cloud services. The concept of direct and recommendation trust is used to compute the comprehensive trust for cloud services. The method for evaluating trustworthiness of a cloud service based on a fuzzy neural network is presented by Wu and Zhou (4). The method also includes a feedback mechanism so that the learning process can be incorporated to classify services into different domains.

The cloud service selection model, which applies the concept of service selected and service delivered, is proposed in Li et al. (5). The method selects an appropriate cloud service based on its trust value. The trust is calculated using different attributes and their assigned weights. The AHP is applied to the calculation of weights. Mukalel and Sridhar (6) have proposed the Trust Management Middleware (TMM) framework to the judge trustworthiness of cloud services. In this chapter, trust is calculated on basis of QoS attributes and user feedback. To guarantee appropriate feedback, it is evaluated by using covariance. The method to reduce trust management overhead is presented in Zhang et al. (7). The paper also presents a method for detecting malicious nodes by applying domain partitioning. The malicious evaluations are detected and omitted from the trust evaluation process by applying the proper filters. The trust evaluation mechanism that uses multiple attributes for the selection of trust levels is proposed in Yang et al. (8). The method utilizes user preferences and AHP for selecting a cloud service. A paper by Hadeel et al. (9) presents a multidimensional trust model to evaluate the trustworthiness of cloud providers. The trust model is applied to big data processing in different clouds. It utilizes the quality of cloud service attributes and user experience with the cloud provider to select a cloud service efficiently. The dynamic and distributed protocol for trust evaluation is proposed by Dou et al. (10). The paper also proposes a method for protecting feedback against any malicious activity and ensuring honest feedback. Homomorphic encryption is applied to the maintain the privacy of feedback. A paper by Wang et al. (11) presents a dynamic trust model that uses SLA parameters to evaluate different cloud services. Here, the comprehensive trust is calculated on the basis of direct trust, indirect trust and reputations. The cloud services are divided into different groups according to their SLA parameters and capabilities. There are many other trust models based on QoS, SLA, reputation, feedback and others are presented in the literature (12–18).

The majority of the existing work applies a static method to assign priority among different QoS attributes and then computing the trust value for a cloud service. However, many of the existing techniques do not combine a machine learning approach with a trust computation for cloud services. Also, there is the need for a dynamic method to decide priority among different considered attributes for trust computation. Hence, the fuzzy c-means clustering based approach, along with user preferences to assign trust levels to cloud service and user both, is proposed here. Compared to existing work, the proposed framework considers a wide variety of attributes to monitor the behavior of a cloud service and a cloud user to enhance the performance.

8.3 PROPOSED METHODOLOGY

Trust is a kind of decision taken by an entity on basis of his own experience and other available knowledge. It is considered as one of the important criteria for selecting an appropriate CSP to deliver a necessary service to cloud users. CSPs also need to evaluate the trustworthiness of cloud users before delivering any service to them. A trust computation framework for cloud services and users is proposed in this chapter. The proposed framework is based on recommendation, user behaviour, SLA parameters and performance. Different types of trust are computed to assess trust levels of cloud services and cloud users. Behaviour trust and recommendation trust are computed for cloud users, while the cloud services are assigned trust levels based on their performance in terms of QoS parameters, their compliance with SLA parameters and recommendation. The proposed model considers that the sensor data are collected and stored in a cloud environment. To guarantee the appropriate utilization and access by trusted entities, the framework for evaluating the trustworthiness of cloud users and services is proposed. The detailed architecture for trust model for cloud services and cloud users is presented in the following subsections.

8.3.1 System Architecture for CSTM_PSR

The system architecture of CSTM_PSR is given in Figure 8.1. The main entities that are involved are WSNs, cloud nodes, cloud users and trust computation framework.
The functionality of each component follows:

1. The data gathered by the sensor nodes in WSNs are shifted to a cloud environment, and the data can be accessed by cloud users.

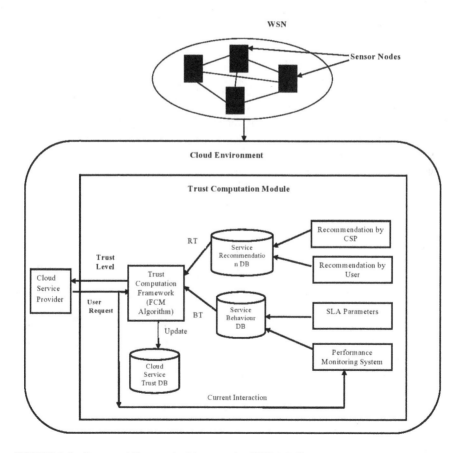

FIGURE 8.1 Proposed System Architecture for CSTM_PSR.

2. Cloud users request any cloud service from the cloud environment that will be handled by the CSP.
3. The CSP will be responsible for verifying the trust levels of a requested cloud service and delivering the efficient service to the cloud user.
4. Trust computation framework: This module is responsible for computation trust levels for all services provided by the cloud nodes. The sub-modules of this framework are as follows:
   a. Service behaviour trust: This module computes trust by consulting the various parameters from performance logs, which are related to QoS and SLA.
   b. Recommendation trust: The feedback collector module collects recommendations from the cloud users and the CPS. The recommendation trust is calculated using these recommendations.
   c. Learning module: The learning module accepts the generated trust values and applies the machine learning algorithm, that is, the fuzzy c-means algorithm, to these values. The algorithm generates different clusters of cloud services based on the computed trust values.
   d. Finally, the trust computation framework selects the trustworthy cloud service for serving the user's request based on the clusters generated in the previous step.

8.3.2 The Trust Evaluation Using CSTM_PSR

The proposed method computes trust levels for cloud services based on performance and recommendation. Performance trust considers different QoS attributes and SLA parameters to measure how trustworthy the cloud service is. Performance trust represents how efficiently and quickly the service has been delivered to the user. The QoS attributes considered for computation of trust include response time, throughput, reliability, availability and success rate. It is the job of the performance monitoring system to record these attributes. On basis of the maintained attribute values, the following evaluation matrix E is generated as given in Equation 8.1:

$$E = \begin{matrix} se11 & se12\cdots & se1m \\ se21 & se22\cdots & se2m \\ sen1 & sen2\cdots & senm \end{matrix} \tag{8.1}$$

Here, each term represents the value of an attribute at different interactions; for example, se_{ij} is the value of jth parameter at the time of ith interaction. The metric of measure is different for every attribute, so it is necessary to perform normalization and bring all attribute values in the same range. The proposed method applies the min–max normalization method, and the corresponding matrix, E_{norm}, that includes all normalized attribute values within range (0–1) is obtained. The obtained normalized evaluation matrix is given in Equation 8.2:

$$E_{norm} = \begin{matrix} s'e11 & s'e12\cdots & s'e1m \\ s'e21 & s'e22\cdots & s'e2m \\ s'en1 & s'en2\cdots & s'enm \end{matrix} \tag{8.2}$$

The user negotiates some security parameters as part of an SLA with the CSP. The trustworthiness of a cloud node can be judged on basis of whether all agreed-to security policies are followed by the cloud node for all interactions. The considered parameters include authentication/authorization, availability of antivirus, encryption method, integrity check and any vulnerability detected. After every interaction, a log is maintained about each security policy attribute.

Let SP = {p1, p2, . . . , pm} is the set of m considered parameters for the SLA. The value either 0 or 1 is assigned to each parameter depending on whether the corresponding security service is provided or not, as shown in Equation 8.3:

$$pi = \begin{cases} 1 \, if \, pi \, is \, provided \\ 0 \, if \, pi \, not \, provided \end{cases}, \tag{8.3}$$

for all i =1 to m.

The obtained evaluation matrix S for these security parameters is given in Equation 8.4:

$$S = \begin{matrix} s11 & s12\cdots & s1m \\ s21 & s22\cdots & s2m \\ sn1 & sn2\cdots & snm \end{matrix} \tag{8.4}$$

The next step is to decide relative importance among the selected attributes and the fuzzy AHP (19) is used for the same. The fuzzy AHP method embeds the fuzzy theory into the basic AHP. The AHP is the most commonly applied method to decision problems based on multiple criteria. It generates the pairwise comparison matrix for the different criteria and develops weights to decide priority among these criteria. The fuzzy logic approach is incorporated into the basic AHP to support vagueness. In the proposed system, the triangular fuzzy function is used to represent the pairwise comparisons of different attributes and to compute the weights for the respective attributes.

The weight matrix, W, is computed using the fuzzy AHP method. Also, the preferences of users are considered when deciding the weights. At the time of the initial agreement with the service provider, the user gives their preference for the QoS and SLA parameters. According to the user's preference, the scale in the pairwise comparison matrix is decided, and the fuzzy weights are obtained. The cumulative performance trust is computed using Equation 8.5:

$$T_{per} = \alpha * (\sum_{i=1}^{n} wi * s'ei) + (1-\alpha) * \sum_{j=1}^{m} wj * s'j. \tag{8.5}$$

Here, T_{per} represents performance trust, w_i is the weight for ith QoS attribute, $s'e_i$ is the evaluation value of ith attribute, w_j is the weight for j^{th} SLA attribute and s'_j is the evaluation value of jth attribute. The weights assigned for each QoS and SLA attribute are denoted by α and $(1 - \alpha)$, respectively.

The recommendation trust is computed on basis of ratings given by each user after interaction and the service provider in the cloud environment. In order to avoid false recommendations, recommendations by highly trusted users and providers are considered. These entities are selected by the trust levels assigned to them in previous *m* interactions. Thus, the recommendations of only the top *m* trusted cloud providers are considered for evaluation, and they are represented in Equation 8.6:

$$T_{recj}^{cs} = \sum_{j=1}^{m} rcs. \tag{8.6}$$

Here, T_{recj}^{cs} is the recommendation trust for the cloud service *j* by the CSP and r_{cs} is the recommendation value obtained based on the previous *m* interactions by the recommender *cs*. Also, the recommendations by the selected users are considered. The users who have interacted with a service continuously are selected to give a recommendation. The user recommendation trust value is computed as shown in Equation 8.7:

$$T_{recj}^{u} = \sum_{j=1}^{m} ru. \tag{8.7}$$

Here, T_{recj}^{u} is the recommendation trust for the cloud service *j* by the user *u* and *ru* is the recommendation value obtained based on the previous *m* interactions by the recommender *u*. The total recommendation trust for the cloud service *j* is represented as T_{recj}, and it is evaluated on basis of recommendation trust obtained from selected recommenders as given in Equation 8.8:

$$T_{recj} = \alpha * \left(T_{recj}^{cs} \right) + (1 - \alpha) * T_{recj}^{u}. \tag{8.8}$$

Here, α $(0 \leq \alpha \leq 1)$ is the weight assigned to the recommendation trust by users and CSPs in the system.

8.3.3 Functioning Scheme: Clustering Cloud Services Using Fuzzy c-Means Method

The fuzzy c-means algorithm is applied to generate the clusters of cloud services. The distance between cluster center and a particular data point is considered and the data point is assigned to a particular cluster depending upon this distance. The updation in the membership and cluster centers is performed after every iteration. Consider that the finite set of data points is X = {x1, x2, ... , xn} and the finite set of clusters is V = {v1, v2, ... vk}. The objective function of algorithm is given by Equation 8.9 (20):

$$J(U,V) = \sum_{i=1}^{n} \sum_{j=1}^{k} (\mu ij)^{m} * \|xi - vj\|^{2}, \tag{8.9}$$

where, $\|x_i - v_j\|$ is the Euclidean distance between j^{th} cluster center and i^{th} data point. U is the fuzzy classification of X, and m is the fuzzification index. The membership value of ith data and j^{th} cluster center are represented by μij. The objective of the algorithm is to minimize Equation 8.9. The cluster centers are calculated using Equation 8.10:

$$avj = \frac{\sum_{i=1}^{n}(\mu ij)^{m} * xi}{\sum_{i=1}^{n}(\mu ij)^{m}} \quad \forall j = 1,2,....k \;, \qquad (8.10)$$

The membership value is computed using Equation 8.11:

$$q\mu ij = \frac{1}{\sum_{c=1}^{k}\left(\dfrac{dij}{dic}\right)^{\left(\frac{2}{m}-1\right)}} \cdot \qquad (8.11)$$

The algorithm that computes the different clusters using previously mentioned formulae and assigns a trust level to each cloud service is given in Algorithm 1.

Algorithm 1: Clustering Cloud Services to Assign Trust Levels (CSTM_PSR)
Input: X = {x1,x2, ... xn} finite data set in n dimensional set,
        k: number of clusters
        θ: threshold for number of iterations
        m: fuzzification index
        $\in$ = termination criterion
        TL={low, medium, avg, high}: finite set of trust levels
Output: Clustering Result C
Trust levels'={x'1,x'2. . . . . , x'n}
Procedure: Fuzzy_Cluster (X, k, m, $\in$, θ)
1.   C = ¢
2.   k = 4;
3.1.1 $\theta = 100$
3.   m = 2
4. $\in$ = 0.01
5.   Compute U = $\left[\mu ij\right]$ the initial membership matrix.
6.   *For i = 1 to θ*
      6.1 Compute cluster centers V = {v1,v2,v3,v4}
      6.2 Update membership function U
      6.3 Compute error = $\mu_{k+1} - \mu_k$ //difference between new and previous centers.
      6.4 If error ≤ $\in$,
          6.4.1 go to Step 7;
          else
          6.4.2 stop.

       End
7.   Plot the clusters C in k dimensional space.
8.   For i =1 to k

$$1.1 \quad avg[i] = \left(\sum_{l=1}^{m} Cl \right) / m \quad //\text{take average of cluster center}$$

    End for
9.   Sort the list of *avg* in ascending order.
10.   For i = i to k   // repeat for all clusters.
      3.2   For each data point x_i in Cluster Cj
          3.2.1 $x_{ij}^{t} = TL[j]$
          End For
      End For

TABLE 8.1
Nomenclature Used in Proposed System

Notation/Acronym	Full Name
CSP	Cloud service provider
QoS	Quality of service
SLA	Service-level agreement
FCM	Fuzzy c-means
X={x1,x2, ... xn}	Finite data set for cloud services in n dimensional set
K	Number of clusters
Θ	Threshold for number of iterations
M	Fuzzification index
∈	Fuzzy c-Means termination criterion
TL	Trust level for cloud services
U	Membership function
C	Cluster center
UTL	User trust level
Y={y1, y2, ... , yn}	Finite data set for cloud users in n dimensional set
T_{per}	Performance trust for cloud service
Tr_{ecj}	Recommendation trust for cloud service
UT_{bev}	Behavioral trust for cloud user
$UT_{rec}j$	Recommendation trust for cloud user
TMM	Trust management module
KNN	k-nearest neighbour
PT	Performance trust
ST	Security parameter–based trust
RT	Recommendation trust
BT	Behavioral trust
CT	Cumulative trust
AHP	Analytical hierarchical process

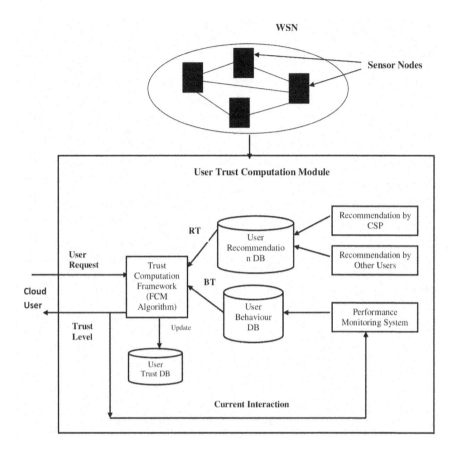

FIGURE 8.2 Proposed System Architecture for CUTM_BR.

End Procedure

The nomenclature used by the proposed system is listed in Table 8.1.

8.3.4 System Architecture for CUTM_BR

The system architecture of the proposed CUTM_BR is given in Figure 8.2. Similar to the trust model for cloud service, here, cloud user trust value is evaluated based on user behaviour and recommendations given to the user. The system architecture includes WSN, cloud user, trust computation framework, user performance monitoring system and recommendation collection system as major components. The CSP transfers the user's request to the trust computation framework so that the trustworthiness of the user can be evaluated before issuing the requested service.

8.3.5 The Trust Evaluation Using CUTM_BR

The trust level of user is computed on basis of their behaviour in the system and recommendations submitted by other entities in the system. To calculate the behaviour trust, the various parameters related to how a user interacts with the system are considered. These parameters include the number of unsuccessful login, caused harm to system, leaked data and upload and download behaviour. The performance monitoring module maintains these parameters for every interaction of the user with the system. The behaviour evidence of all considered attributes is maintained in the form of a matrix as given in Equation 8.12:

$$UE = \begin{matrix} ue11 & ue12 & \ldots & ue1m \\ ue21 & ue22 & \ldots & ue2m. \\ uen1 & uen2 & \ldots & uenm \end{matrix} \tag{8.12}$$

Here, ue_{ij} represents the value of jth attribute during the ith interaction.

The appropriate normalization method is applied to matrix in Equation 8.12 to ensure that all parameters lie in the range (0–1). The updated normalized matrix is given in Equation 8.13:

$$UE_{norm} = \begin{matrix} u'e11 & u'e12 & \ldots & u'e1m \\ u'e21 & u'e22 & \ldots & u'e2m. \\ u'en1 & u'en2 & \ldots & u'enm \end{matrix} \tag{8.13}$$

As discussed in the previous section, it is necessary to assign priorities to the selected attributes. Here also, the fuzzy AHP method is used to assign a relative importance scale to the attributes. The fuzzy weight matrix W is computed using the fuzzy AHP method. The user behaviour trust is computed using Equation 8.14:

$$UT_{bev} = \left(\sum_{i=1}^{n} wi * u'ei \right). \tag{8.14}$$

Here, UT_{bev} represents behaviour trust, w_i is the weight for i^{th} attribute and $u'e_i$ is the evaluation value of ith attribute.

The recommendation trust is computed on basis of ratings given by the other users and service providers in the system. The recommendations by trusted users are only considered. These users are selected on basis of their previous interactions, and the user trust database is also consulted. The total recommendation trust for the cloud user j is represented as UT_{recj}, and it is evaluated on basis of recommendation trust obtained from selected users and CSPs by using Equation 8.15:

$$UT_{recj} = \beta * \left(\sum_{j=1}^{m} rcsp \right) + (1-\beta) * \left(\sum_{j=1}^{m} ru \right). \tag{8.15}$$

Here, β $(0 \le \beta \le 1)$ is the weight assigned to the recommendation trust by other users and CSPs in the system.

8.3.6 FUNCTIONING SCHEME: CLUSTERING USING THE FUZZY C-MEANS METHOD

The algorithm that computes the different clusters using a fuzzy c-means algorithm discussed in the previous section and assigns a trust level to each cloud user is given in Algorithm 2.

Algorithm 2: Clustering Cloud Users to Assign Trust Levels (CUTM_BR)

Input: Y={y1,y2, ..., yn} finite data set in n dimensional set

       k: number of clusters

      θ : threshold for number of iterations

      m: fuzzification index

      $\in$= termination criterion

      UTL={low, medium, avg, high}: finite set of trust levels

Output: Clustering result C

Trust levels {y′1,y′2. . . . . y′n}

Procedure: Fuzzy_Cluster(Y, k, m, $\in$, θ)

1.   C = ¢
2.   k = 4
3.2.2 θ = 100
3.   m = 2
4. $\in$ = 0.01
5.   Compute U = $\left[\mu ij\right]$ the initial membership matrix.
6.   *For i = 1 to* θ
      6.1   Compute cluster centers UV={uv1,uv2,uv3,uv4}
      6.2   Update membership function U
      6.3   Compute error = $\mu_{k+1} - \mu_k$ //difference between new and previous centers.
      6.4   If error ≤ $\in$
          6.4.1 go to step 7;
          else
          6.4.2 stop.
          End
7.   Plot the clusters C in k dimensional space.
8.   For i = 1 to k

$$1.2 \quad avg[i] = \left(\sum_{l=1}^{m} Cl\right) / m \quad \text{//take average of cluster center}$$

    End for
9.   Sort the list of *avg* in ascending order.
10. For i = i to k   // repeat for all clusters
      3.3   For each data point x$_i$ in cluster Cj
          3.3.1 $y_{ij}^t = UTL\left[j\right]$
          End For
      End For
End Procedure

8.4 SIMULATION

This section elaborates the evaluation of the proposed work with respect to different performance parameters and a comparison with some relevant existing schemes. The proposed method prototype is implemented in Java, and the Simulink simulator is also used. It is considered as the data derived by sensor nodes are already shifted to the cloud, and the proposed method calculated the trustworthiness of users and services processing these data. The trust levels are computed for cloud resources and cloud users by applying the algorithms described in Section 8.3. The different experiments are performed a number of times, and the trust levels of cloud users and services are measured. The experiments are conducted to compare the performance of the proposed algorithm with the TMM model presented in Khilar et al. (2). Trust evaluation is completed using a k-means clustering method and trust evaluation with fuzzy AHP method.

8.4.1 SIMULATION MODEL

The different trust levels considered are low, average, medium and high. The trust levels for cloud services are computed using QoS attributes SLA parameters and recommendations. The WSDREAM (21) data set is used for a few QoS attributes, such as response time and throughput. For the remaining QoS attributes, a synthetic data set is created by following a similar approach. The SLA parameters data set is also randomly generated for the considered parameters. In this data set, the value assigned for a particular parameter can be either 0 or 1, depending on the absence or presence of it in the SLA. Similarly, the synthetic data set for recommendations by CSPs and other users is generated. Here, the CSPs or users can provide ratings in the range (1–10). The recommendations are recorded based on the previous 10 interactions with the particular cloud service. The final rating is the average of ratings generated in last 10 interactions. A similar approach is followed for other users' ratings. The six different data sets for various attributes of performance, security and recommendation are generated with 100, 500, 1000, 5000 and 10,000 interactions, respectively. Initially, the medium trust level is assigned to all cloud services. The chapter also presents cloud users' trust evaluation strategy. The cloud user behaviour attributes and recommendations by CSPs and other users are considered for this purpose. The required data set for the behaviour attributes and recommendations are generated randomly. The behaviour data set can contain values in the range (1–10). These values indicate how many times the specific attribute is satisfied for the user in the previous 10 interactions with the cloud environment. The recommendation data set is created by following a similar approach applied for cloud service recommendations. Here also, six different data sets with varying numbers of interactions are generated to compare the performance of the proposed system.

The proposed algorithm is applied to the designed data sets at different time intervals. The proposed algorithm is applied to group the cloud services into four different clusters. The trust level is assigned to each cloud service in a particular cluster based on the cluster centroid so that the trustworthy cloud service can be assigned to the cloud user. At the same time, the cloud users' trustworthiness is evaluated to guarantee that the trusted user is accessing the designated cloud service. In this way, the proposed model ensures the secure delivery of a cloud service to any cloud user by applying the trust evaluation both ways.

8.4.2 Performance Parameters

The parameters defined for evaluating the performance of the proposed model are as follows:

1. Accuracy: The accuracy of the proposed model represents the correctness in classifying the trust values of cloud services and cloud users.
2. Mean Execution Time (MET): Along with accuracy, the MET is also evaluated and compared for different data sets. The MET is measured in terms of how much time is required to generate the cumulative trust value for cloud service or user and assign the trust levels to them.
3. Average Trust: The average trust value of cloud service in terms of performance trust, security parameter–based trust and recommendation trust and cumulative trust for each data set are calculated. This value indicates the effect of a varying number of cloud services on trustworthiness. The user's behaviour trust, recommendation trust and cumulative trust are considered to generate an average trust value.
4. Success Rate: The success rate indicates how many interactions are completed successfully by applying trust levels to cloud services and cloud users and granting access and selecting service on the basis of trust levels.

8.4.3 Results and Discussion

The proposed models CSTM_PSR for cloud services and CUTM_BR for cloud users described in Section 8.3 are compared with three different methods, that is, TMM (2), weighted sum method using fuzzy AHP and trust evaluation using a k-means clustering algorithm. All the mentioned methods are implemented using the simulation environment mentioned earlier in this section. The performance parameters mentioned previously are compared by analysing different iterations done over the generated data sets. The TMM method uses a simple AHP method to calculate the weights for attributes. In this model, different machine learning techniques are presented to compute the trust value. The proposed methods are compared with the results of the KNN algorithm presented in Khilar et al. (2). The k-means clustering algorithm is also implemented for the generated data sets for cloud users and cloud services. In the case of the fuzzy AHP method, the weights for the different considered attributes are computed, and then the weighted sum method is applied to compute the cumulative trust.

Figure 8.3 represents the performance comparison of all four models based on accuracy. The accuracy is measured in terms of how many cloud services are correctly classified according to the trust levels. All data sets are divided into training and testing parts. Of the data set, 70% is utilized for training, and 30% is utilized for testing purposes. The figure indicates that the proposed CSTM_PSR model has highest accuracy for almost all data sets. The fuzzy AHP method indicates the lowest

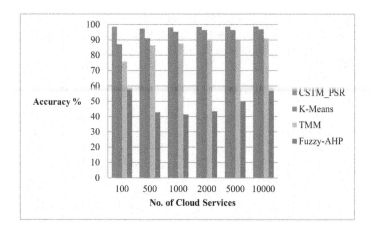

FIGURE 8.3   Comparison of Accuracy for Four Algorithms.

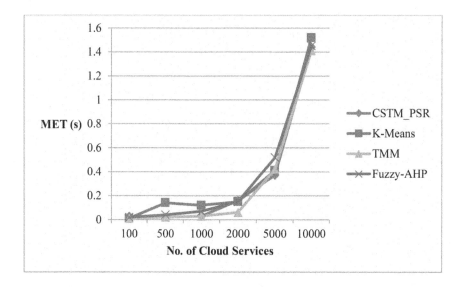

FIGURE 8.4   Comparison of MET for Four Algorithms.

accuracy value for all data sets. Thus, it can be concluded that the integration of a machine learning approach into the traditional trust evaluation methods is advantageous. The figure indicates that the machine learning approach in TMM, the proposed CSTM_PST and the k-means have higher accuracy than the plain fuzzy AHP method. The comparison of execution time required for all algorithms is shown in Figure 8.4. The MET is considered for an evaluation that includes time for training and testing both. The time taken by TMM is less compared to the proposed CSTM_PSR method

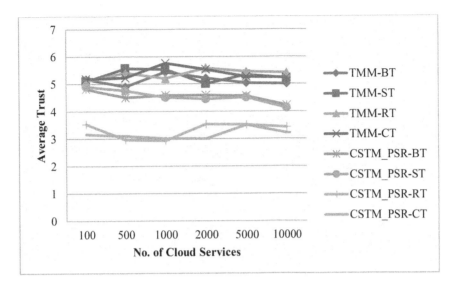

FIGURE 8.5   Comparison of Average Trust Value.

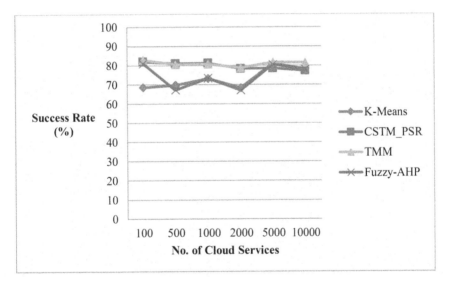

FIGURE 8.6   Comparison of Success Rate of Interaction with Trustworthy Cloud Service.

for a data set with a fewer number of services. But for the largest data set, with 10,000 services, the proposed algorithm outperforms all the other algorithms.

Figure 8.5 shows the comparison of average trust value of CSTM_PSR and TMM method for four types of trust, that is, Performance Trust (PT), Security Parameter–Based Trust (ST), Recommendation Trust (RT) and Cumulative Trust (CT). The comparison is done for all six data sets by performing different iterations of the mentioned algorithms. It can be observed from the figure that the average trust value decreases as the number of cloud services in the system increases. This change in the average trust value for the proposed method is consistent; hence, the accuracy in the trustworthiness evaluation can be high for the proposed model.

The effect of trust evaluation on the success rate of the interaction with cloud environment is depicted in Figure 8.6. The success rate of TMM and the proposed model is almost similar. The success rate goes down as the number of cloud services increase in the cloud environment. But for the k-means method and the fuzzy AHP method, there is no consistency in the value of success rate.

Figure 8.7 shows the comparison of the number of cloud services grouped in four different clusters with Trust Levels (TLs) low, medium, average and high for all data

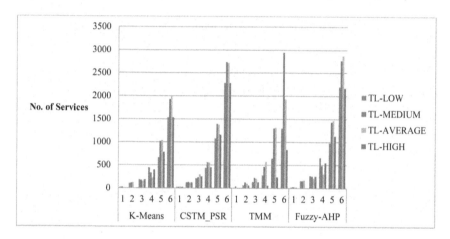

FIGURE 8.7 Comparison of Cluster Assignments by CSTM_PSR.

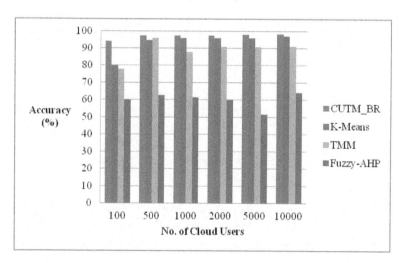

FIGURE 8.8 Comparison of Accuracy for Four Methods.

sets. For the proposed model, the number of services with low TL grows in proportion to the total number of available services. The variation in the number of services in each level is also consistent.

Next, the comparison of the proposed CUTM_BR model that judges the trustworthiness of cloud users with the existing trust models is presented. The performance parameters mentioned in earlier in this section are used for comparison. The proposed model is compared with the TMM (2) model, the fuzzy AHP method and the k-means clustering method. Similar to the proposed CSTM_PSR, this model is

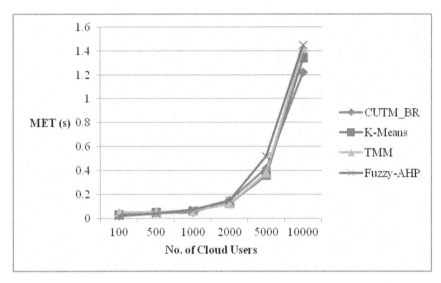

FIGURE 8.9   Comparison of MET for Four Methods.

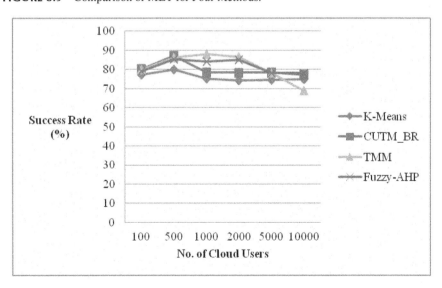

FIGURE 8.10   Comparison of Success Rate of Interaction with Trustworthy Cloud Users.

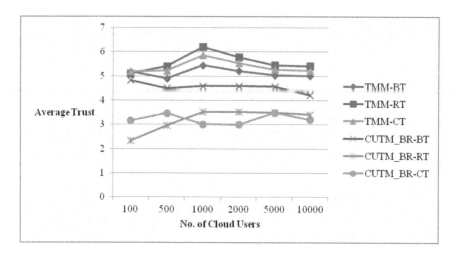

FIGURE 8.11   Comparison of Average Trust Value for CUTM_BR and TMM.

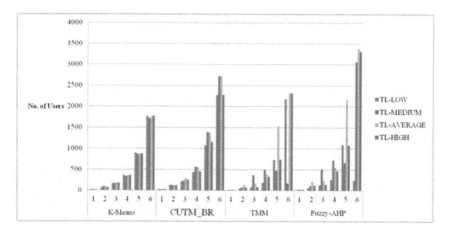

FIGURE 8.12   Comparison of Cluster Assignments by CUTM_BR.

also evaluated on basis of accuracy, MET, average trust value and success rate. The comparison of accuracy of all four algorithms is shown in Figure 8.8. The proposed model outperforms all other models as it gives a high accuracy for all generated data sets. Thus, the proposed CUTM_BR model correctly classifies cloud users into different groups according to their trust levels.

The execution time comparison for all four methods is shown in Figure 8.9. Here also, the proposed model performs better than other models.

The comparison of the number of successful interactions of cloud users with the cloud environment is shown in Figure 8.10. The comparison shows that the success rate is lower when the number of users increases in the system.

Similar effects can be noticed for the average trust value evaluated by the proposed CUTM_BR model and the TMM model. The average trust value decreases as the number of users in the system increases as presented in Figure 8.11.

Figure 8.12 represents the assignment of cloud users into different clusters as per their trust levels. Similar to the CSTM_PSR model, this model also shows the proper allocation of users into different groups.

8.5 CONCLUSION

This chapter presents an integrated approach for WSNs and cloud computing for storing sensor data in a cloud environment. A trust evaluation mechanism for cloud users and services is proposed to ensure that the data are accessed and processed by trusted entities. The chapter presents the two trust models: CSTM_PSR to compute trust levels for cloud services and CUTM_BR to compute trust levels for cloud users. The performance parameters, SLA parameters and recommendations are considered while evaluating the trust levels of cloud services. In the same way, the CUTM_BR model computes trust for cloud users by evaluating users' behaviours and recommendations. Different weights are assigned to the selected attributes on basis of user preferences. Thus, the dynamic preference-based method is applied along with a fuzzy AHP method to compute the different types of trust. The proposed model combines traditional trust computation with machine learning techniques to enhance performance. The fuzzy c-means clustering algorithm is applied to generate clusters of both cloud services and cloud users. Then the trust levels are assigned to the cloud services or cloud users depending on the cluster they are mapped to. The existing TMM method is also implemented to compare the performance of the proposed method. The k-means clustering algorithm and the fuzzy AHP method without any learning ability are also implemented. A synthetic data set with a various number of users and services is generated to test the performance of all algorithms. The experimental results show that the accuracy of the proposed model is better than the other three algorithms. In terms of execution time and success rate, the proposed algorithm outperforms the existing techniques.

In the future, the proposed model can be enhanced by including a wide range of QoS and behavioral attributes. Also, some mechanisms for judging false recommendations can be incorporated into proposed system. The performance of the proposed model can be further evaluated by applying other machine learning or deep learning techniques.

REFERENCES

1. Carlos-Mancilla, Miriam, Ernesto López-Mellado, and Mario Siller. (2016). Wireless Sensor Networks Formation: Approaches and Techniques. *Journal of Sensors*, 2.
2. Khilar, P.M., V. Chaudhari, and R. R. Swain. (2019). Trust-Based Access Control in Cloud Computing Using Machine Learning. In H. Das, R. Barik, H. Dubey, and D. Roy. (eds), *Cloud Computing for Geospatial Big Data Analytics: Studies in Big Data*. New York: Springer, 49.
3. Wang, Yubiao et al. (2019). A Cloud Service Selection Method Based on Trust and User Preference Clustering. *IEEE Access*, doi:10.1109/ACCESS.2019.2934153.

4. Wu, Zhengping, and Yu Zhou. (2016). Customized Cloud Service Trustworthiness Evaluation and Comparison Using Fuzzy Neural Networks. *IEEE 40th Annual Computer Software and Applications Conference*, 433–442.

5. Li, X., H. Liang, and X. Zhang. (2016). Trust Based Service Selection in Cloud Computing Environment. *International Journal of Smart Home*, 10(11), 39–50.

6. Mukalel, B. S., and R. Sridhar. (2019). TMM: Trust Management Middleware for Cloud Service Selection by Prioritization. *Journal of Network System Management*, 27, 66–92.

7. Zhang, Pei Yun, Yang Kong, and Meng Chu Zhou. (2018). A Domain Partition-Based Trust Model for Unreliable Clouds. *IEEE Transactions on Information Forensics and Security*, 13(9), 2167–2178.

8. Yang, Y., R. Liu, Y. Chen, T. Li, and Y. Tang. (2018). Normal Cloud Model-Based Algorithm for Multi-Attribute Trusted Cloud Service Selection. *IEEE Access*, 6, 37644–37652, doi:10.1109/ACCESS.2018.2850050.

9. Hadeel, T. E., A. S. Mohamed, D. Rachida, and B. Boualem. (2018). A Multi-Dimensional Trust Model for Processing Big Data Over Competing Clouds. *IEEE Access*, 6, 39989–40007, doi:10.1109/ACCESS.2018.2856623.

10. Dou, Y., H. C. B. Chan, and M. H. Au. (2019). A Distributed Trust Evaluation Protocol with Privacy Protection for Intercloud. *IEEE Transactions on Parallel and Distributed Systems*, 30(6), 1208–1221, doi:10.1109/TPDS.2018.2883080.

11. Wang, Y. J. Wen, W. Zhou, and F. Luo. (2018). A Novel Dynamic Cloud Service Trust Evaluation Model in Cloud Computing. *17th IEEE International Conference on Trust, Security and Privacy in Computing and Communication*, 10–15, doi:10.1109/TrustCom/BigDataSE.2018.00012.

12. Wu, X. (2018). Study on Trust Model for Multi-Users in Cloud Computing. *International Journal of Network Security*, 20(4), 674–682.

13. Udaykumar, S., and T. Latha. (2017). Trusted Computing Model with Attestation to Assure Security for Software Services in a Cloud Environment. *International Journal of Intelligent Engineering & Systems*, 10(1), 144–153.

14. Jagpreet, S., and S. Sarbjeet. (2017). Improved Topsis Method Based Trust Evaluation Framework for Determining Trustworthiness of Cloud Service Providers. *Journal of Grid Computing*, 15, 81–105.

15. Shilpa, D., and I. Rajesh. (2017). Evidence Based Trust Estimation Model for Cloud Computing Services. *International Journal of Network Security*, 20(2), 291–303.

16. Challagidad, P. S., and M. N. Birje. (2020). Multi-Dimensional Dynamic Trust Evaluation Scheme for Cloud Environment. *Journal of Computer & Security*, 91.

17. Yubiao, W., W. Junhao, W. Xibin, T. Bamei, and Z. Wei. (2019). A Cloud Service Trust Evaluation Model Based on Combining Weights and Gray Correlation Analysis. *Journal of Security and Communication Networks*, 1–12.

18. Yiqin, L., F. Yahui, and Q. Jiancheng. (2019). A Trust Assessment Model Based on Recommendation and Dynamic Self-Adaptive in Cloud Service. *Journal of Physics: Conference Series*, 1325, 1–7.

19. Demirel, T., N. Ç. Demirel, and C. Kahraman. (2008). Fuzzy Analytic Hierarchy Process and Its Application. In C. Kahraman. (eds), *Fuzzy Multi-Criteria Decision Making. Springer Optimization and Its Applications*. New York: Springer, 16, doi:10.1007/978-0-387-76813-7_3.

20. *Fuzzy c-Means Clustering Algorithm*, https://sites.google.com/site/dataclustering algorithms/fuzzy-c-means-clustering-algorithm.

21. Zhang, Y., Z. Zheng, and M. R. Lyu. (2011). Wspred: A Time-Aware Personalized QOS Prediction Framework for Web Services. *22nd International Symposium on Software Reliability Eengineering (IS-SRE'11)*, 210–219.

9 Design of Wireless Sensor Networks Using Fog Computing for the Optimal Provisioning of Analytics as a Service

Rajalakshmi Shenbaga Moorthy and P. Pabitha

CONTENTS

Recently, the Internet of Things (IoT), together with wireless sensor networks (WSNs), generates a huge volume of data that necessarily needs to be processed in real time. Currently, the data to be processed are sent to the cloud, which is not fit for lower-latency applications. Although cloud computing provides access to data from anywhere and at any time, moving such a huge volume of data to a cloud incurs large time requirements, energy consumption, and cost and affects throughput. As the cloud is located physically in a distant location, it is difficult to process

DOI: 10.1201/9781003107477-9

IoT applications with appropriate latency and throughput. Analytic challenges arise when large amounts of data originated from IoT devices is scheduled for Analytics as a Service (AaaS) providers. When the data grow exponentially, the analytic engine at the cloud faces challenges such as increased processing time, processing cost, energy consumption, and latency in processing the large analytic requests. Such delays in processing are often critical for time-sensitive applications such as health care, which results in non-optimal insights. To overcome these challenges, the analytic requests are processed by the proposed fog engine, which is placed in the fog layer. Also, to do optimal analytics at the fog engine, a metaheuristic clustering algorithm based on the Firefly algorithm (FFA) and fuzzy k-means (FKM) is proposed and is simulated in Java, and the results show that analyzing data at a fog engine offers significant advantages in terms of latency, energy consumption and network usage time.

9.1 INTRODUCTION

Cloud computing is pervasive and ubiquitous computing that attracts a wide range of users as per the slogan pay per use by providing a virtualized pool of computing resources, on-demand network access, and large storage capacity (1, 51). These features made cloud computing more attractive in the technological era. However, recent emergence in the field of IoT, which includes fields ranging from smart cities to pervasive health care, brings challenges to cloud computing, especially by means of high communication overhead, data privacy (2), unpredictable latency, a paucity of mobility support, and location awareness (3). Fog computing was designed to overcome some of the problems of the cloud by providing data privacy, security, business agility (4), and services to the user at the edge of computing (3). Fog computing is heterogeneous as it is situated on the border of the internet. Fog computing includes a collection of fog nodes plugged with cores that have a high processing speed along with in-built storage capacity. It is quite difficult to establish a seamless connection between fog nodes and offering services over that in the era of the IoT. Thus, techniques such as software-defined networking (5) have emerged to create a pliable networking environment (3, 7).

Fog computing serves the need of users within a limited coverage, and the fog server can directly interact with users within the range. Also, fog computing provides local services, and fog nodes are distributed (6). Although the fog and the cloud use the same mechanisms, such as computing, storage, and network, along with the other means such as virtualization and multi-tenancy, fog computing handles time-sensitive applications better than cloud computing (9). Fog computing devises a model in which data generated from IoT devices are analyzed by the application present in the device rather than in a centralized cloud server. The feature that made fog computing more attractive is that it can handle a situation in which the number of devices in the network keeps increasing (10). If all the devices are connected to the internet and produce data, then, at a certain point, it is not worth transferring data to the cloud. At that time, when the data get transferred to the cloud for analyzing, the data may be of no use. To handle this situation, the gateway device that connects to an IoT device and cloud must be programmed with extra functionality for analyzing the data before transferring. This way of communication avoids network congestion (11). Fog computing is an addition to cloud computing, with additional characteristics such as scalability, wireless access, handling real time-sensitive applications (12). Thus, fog computing serves as an idle platform for handling time-sensitive applications that require immediate analytics. Some of the examples that require outcome at once, that is, low latency, are smart traffic light systems (9), wind farms (9), and e-health care (8).

Heart disease prediction is essential in today's pervasive world (41). Latency in predicting heart disease may have a serious impact on human life. Thus, in this chapter, distributed analytics as a service is proposed using a fog engine for a heart disease data set taken from University of California, Irvine (UCI) repository to overcome the challenges arising in analytics using centralized cloud computing. A wide range of machine learning algorithms are available for predicting heart disease. To optimally predicting heart disease, that is, maximizing the accuracy of the classifier, a metaheuristic clustering is introduced by integrating a metaheuristic algorithm called FFA with FKM. The proposed metaheuristic clustering is plugged in a fog engine for optimal analytics.

The rest of the chapter is organized as follows: the motivation of distributed analytics as a service is described in Section 9.2. Some of the existing applications that utilize fog computing is described in Section 9.3. Sections 9.4 and 9.5 describe the proposed tree-structured topology for distributed analytics as a service. Section 9.6 describes the implementation of metaheuristic clustering algorithm in the fog engine. Experimental results are described in Section 9.7, and Section 9.8 concludes the work with future impacts.

9.2 MOTIVATION

With a huge number of sensors, IoT devices generate an unrivaled amount of data, which, when carefully analyzed, produce better insights for decision-making in various fields. When a huge amount of data gets generated, a situation called the analytics imperative arises (50). Time-sensitive applications like health care or applications that require less latency incur a significant drawback when processing data in the cloud.

As a result, there is a need for

- a topology/framework for distributed processing of analytic requests that are really latency-sensitive and
- proposing distributed data analytics near the user to minimize latency, network usage time, energy consumption, processing time, and processing cost and to fasten response time.

Thus, the fog engine, which is present very close to IoT devices and analyzes the data, was proposed. This avoids congestion and provides faster decision-making. The fog engine lies between cloud resources and the IoT is needed to efficiently analyze the user's data with minimum delay (8). The main contributions of this chapter are as follow:

- A tree-structured topology for distributed analytics as a service
- Data analytics happening at the place where it is generated through the fog engine

- A fog engine-based distributed analytics as a service using a metaheuristic clustering technique
- An attempt to implement a metaheuristic clustering technique in a fog engine
- Application of the proposed technique to a data set from the UCI repository (42) and the metrics such as latency, energy consumption, and network usage time analyzed

Figure 9.1 shows the reason behind proposing a fog engine for processing analytic requests. Figure 9.1a refers to the processing of a user's request in cloud computing, which takes a longer communication time and response time and took more bandwidth. When the user wishes to see whether they will be subject to a particular disease, the user sends the symptoms to the hospital, which runs the predictive model using cloud computing and thus forwards the user's request to the cloud for processing. After processing, the outcome is delivered to a hospital, and the hospital sends the outcome to the intended patient. Thus, the process always involves communication overhead. The problem of communication overhead is solved by fog computing, represented in Figure 9.1b. In Figure 9.1b, the user sends their symptoms to the hospital, where the fog server has a prediction model. Having done the analytics at the fog server, it immediately responds to the user about the outcome of the prediction. Thus, from Figure 9.1, it is observed that fog computing responds much faster than cloud computing and with reduced latency. The response time for processing the user's request is also less, which has a significant reduction in the required bandwidth for communication. Also, the user can enjoy the benefit of very lower application service costs (6).

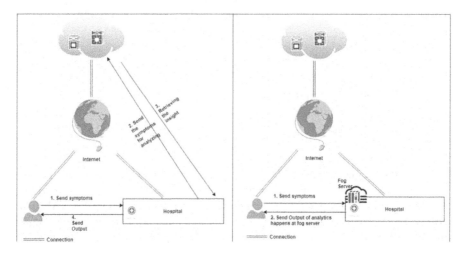

FIGURE 9.1   (a) The Process of Analytics in Cloud Computing, (b) the Process of Analytics in Fog Computing.

9.3 RELATED WORK

Harishchandra Dubey et al. designed an architecture called fog data to perform analytics on the data gathered from IoT devices for pervasive health care. It includes three subsystems for carrying analytics and a cloud server for offline processing. The dynamic time wrapping (DTW) algorithm is used to create patterns using time-sensitive information gathered from wearable sensors. The algorithm works by computing the similarity between two tuples, even though they are of different lengths. The clinical speech processing chain (CLIP) algorithm to recognize speech, pan Tompkins for recognizing speech, and DTW were implemented in fog nodes to reduce bandwidth usage (13). Rabindra K. Barik et al. developed a fog geographic information system (GIS) to analyze geographical data. The FogGIS framework aimed to raise outcomes while transmitting and analyzing social data. The fog computing device in the FogGIS framework performs overlay analytics. Overlay analysis is one in which various geospatial data are superimposed in a unique platform to analyze raster and vector of geospatial data. The overlay analysis was carried for the city in Alaska (14). Hamid Reza Arkian et al. designed a fog-based data analytics system to minimize the cost for provisioning of resources for IoT crowdsensing applications (15). Ala Al-Fuqaha et al. extensively studied the key ingredients of IoT and its components. The IoT, which can integrate a billion or a trillion physical objects through the internet, does not have a common underlying architecture. Various architectures for the IoT include a three-layered one that consists of the application layer, network layer, and perception layer; middleware-based architecture; service-oriented architecture-based architecture and five-layered architecture. To strengthen data analytics, using data generated from various IoT devices, there must be a standard architecture that better use the protocols and components of the IoT (16).

Prem Prakash Jayaram et al. designed CARDAP (Context-Aware Distributed Analytic Platform). For applications that are not time-sensitive, data get stored in the platform for later processing. The query latency cost model was designed to minimize the cost required for the time taken to answer the analytic request (17). Shree Krishna Sharma et al. designed a live data analytics system in wireless IoT networks by integrating cloud and fog computing. In the collaborative edge cloud architecture, IoT gateways have cache memory and can deliver the cached content even in offline mode. The edge devices can be anything that has the capability to store compute and is plugged with seamless networking functionality. IoT devices are heterogeneous and generate a vast amount of data with a large volume of features. The analytic request that is time- and delay-sensitive is processed at the edges. The analytics are postponed in the cloud for other requests. The framework also supports using the data stored in the cloud for deciding on the edges without holding back for data streaming from IoT devices (18). Khubaib Amjad Alam et al. designed far-edge analytics in mobile devices. Thus, far-edge analytics has potential benefits such as the reduction in data in a mobile cloud computing environment (MCCE) and a reduction in the rate of transferring data and the cost for utilization of bandwidth. The frequent pattern mining algorithm is executed in mobile devices, which considerably reduces the movement of data between the mobile device and the cloud. To handle a diverse set of data, chunked data analysis, context-based knowledge discovery, and periodic data analysis are used for processing the data (19).

Chamil Kulathunga et al. designed smart dairy farming using a delay-tolerant networking (DTN) algorithm. DTN aims to transfer data as bundles between source to destination. If no direct connection is available, data will be directed via an intermediate path. Delay-critical data, like the health condition of animals, which requires immediate analysis, are processed in edge devices (20). Kuljeet Kaur et al. discuss the challenge of high bandwidth consumption and the long latency period taken for processing a huge volume of instantaneous data from the IoT in cloud. Edge computing was coined to overcome the previously mentioned challenge and provides facilities for processing expeditious data at the edge device itself. The problem of job scheduling with multiple objectives, such as minimizing energy consumption and makespan using cooperative game theory, was addressed with a container-based edge service management system (CoESMS) (21). Jianhua He et al. designed a large-scale data analytics system using ad hoc and dedicated fogs. Resource management for scheduling analytic jobs to satisfy quality of service (QoS) was designed using a multitier fog computing system (22).

Shusen Yang et al. addressed the processing of low-latency applications, such as video mining and event monitoring, in fog computing. The streaming data created by IoT devices, such as microphones and camera, were collected at the fog server, such as WSNs and cellular networks, and the gathered data were processed to deliver insights. The insights were also stored in the fog server for offline processing. The architecture included five layers, such as the application layer, the processing layer, the data management layer, the resource management layer, and the physical layer. The architecture considered streaming algorithms for processing data, the privacy of data, pricing of services, and an optimization algorithm for better resource allocation (23). Hazeem M. Raafat et al. designed a feature extraction method to extract relevant features from the input sensor data and fed into neural networks for classification (24). Pankesh Patel et al. summarizes the challenges in implementing fog architecture such as a dynamic application framework for analyzing streaming IoT data in contrast with static frameworks such as Hadoop. Furthermore, fog computing handles a large number of heterogeneous devices than cloud computing does and thus requires heterogeneous protocols and hardware to handle diverse sets of operations. The deployment of heterogeneous fog nodes has to be taken care of when creating a fog network (25).

Rajat Chaudhary et al. proposed a methodology, whereby the network is built using software-defined networking and network functions virtualization. The integration brings various security issues such as a distributed denial-of-service (DDoS) attack, which can be protected using the Kerberos authentication server. The integration is required to deal with the mobility of data, data storage, and analytics (26). Hesham El-Sayad et al. give insight into edge computing, fog computing, and cloud computing. Furthermore, the performance of edge computing is analyzed in various networking environment. Edge computing, where the processing is done at the edges, uses a radio access network (RAN) for communication. Fog computing has a similar capability as that of the edge, and both do analytics at closer proximity to the end user. But the former uses Local Area Network (LAN) for communication, and the latter uses RAN for communication. Cloud computing does analytics at the centralized cloud server that is too far from the user and thus introduces long delay and high response times (27). Tasneem S.J. Darwish et al. designed an architecture to handle a transportation model by integrating three dimensions. The lambda architecture for large-scale analytics was proposed in which data are grouped as a view and the query is searched on the view. There are three layers to handle data analytics, namely, a batch layer, where the views will be created from the collected data; a speed layer, where it enables parallel processing; and a serving layer, which handles both ad hoc requests and streaming application requests (28).

Sherif Abdelwahab et al. devised FogMQ, an application in edge and fog computing. The end-to-end latency for IoT applications increases because of multi-hop queuing models and brokers exist between the source to the destination. Autonomous clone migration protocol is used to reduce latency and to maximize faster response time between peers. Tomography functionality is used by the FogMQ server to carry out the migration decision of analytics without the need to worry about hosting platforms (29). Xiaolong Xu et al. designed an oriented track, a planning system, to help people in disasters. The system is built on IoT/fog computing using an artificial potential field with relationship attraction (APF-RA). The various attributes taken into account to rescue people in danger are route length, shifting time, and convergence rate (30). Abebe Abeshu Diro et al. proposed a lightweight encryption mechanism called elliptic curve cryptography (ECC)–based proxy re-encryption. A lightweight security mechanism can be achieved by implementing security algorithms in fog rather on the IoT. The designed ECC-based proxy re-encryption achieves better results in terms of encryption and decryption runtime and throughput (31). Gaocheng Liu et al. addressed the problem of object tracking, such as degrading performance when illumination gets varied and the speed of the object. A framework for intelligent video surveillance system was built on the fog computing environment. Multi-position detection and use of an alternate template and correlation filter-based tracker were implemented to track objects. Although the performance of the proposed object tracking is better, the other challenges in fog computing, such as resource management, security, and privacy, have to be viewed seriously (32). Diana C. Yacchirema et al. provided a solution for obtrusive sleep apnea (OSA). The immediate solution for this disorder is required for elderly people. The factors taken into account for the prediction of OSA are sleeping environment and sleeping status. An artificial neural network with a ReLU, rectified linear activation function, was used for analyzing the data. The fog layer gets the current status of an elderly person, does analyze, and reports the insights to caretakers at once. On the other hand, descriptive analytics is done at the cloud layer using batch data generated from the fog layer, thereby promoting quality of life (33). Pradip Kumar Sharma et al. proposed the SoftEdgeNet model based on software-defined networking with blockchain for edge computing to overcome challenges such as bandwidth limitation, latency, and real-time analysis for a vast amount of data generated from IoT devices. The model is decentralized and consists of five layers, such as data producers, consumers, an edge layer, a fog layer, and a cloud layer. The data from IoT devices are forwarded to the edge layer for immediate processing. The insights are forwarded to consumers in addition to the fog layer. The fog layer is responsible for detecting any dangerous events and reporting it to the edge layer (34). Razi Iqbal et al. proposed a data analytic framework for a fog computing environment to handle data generated by the Internet of Vehicles (IoV). The challenges of IoV are heterogeneity, interoperability, real-time data, QoS, context awareness, and data quality. The proposed framework includes a data collection layer, a preprocessing layer, an analytics layer, a services layer, and an application layer that can handle real-time analytics, near-real-time analytics, batch analytics, and context reasoning (35). Yu Wang et al. proposed a disaster area adaptive delay minimization (DAADM) to minimize delay in the emergency communication network. The DAADM is fused in an analytics module of the fog layer. The proposed adaptive algorithm is compared with the genetic algorithm, and the results are proof that it is best suited for real-time applications (36). Mohit Taneja et al. (37) proposed the decomposition of multivariate linear regression for data analytics to avoid transferring raw data to centralized architecture which incurs delay.

From the literature review, the authors found that most of the researchers intended to design a novel architecture in fog computing to minimize bandwidth utilization, latency, and response time. Only a few researchers focused on machine learning algorithms for performing analytics at the fog layer. The integration of a metaheuristic algorithm and machine learning algorithms at the fog layer is done, although such approaches already exist at the cloud computing level (38, 39, 40, 45, 47). The general three-tier architecture of fog computing for performing data analytics is specified in Figure 9.2. Table 9.1 summarizes the key comparison of cloud and fog computing.

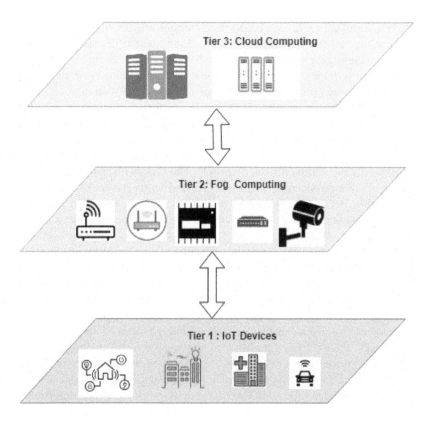

FIGURE 9.2   Three-Tier Architecture for Fog-Based Distributed Analytics as a Service.

TABLE 9.1

Differences of Key Features of Cloud and Fog Computing

Features	Cloud	Fog
Devices	High-End servers	IoT devices, routers, gateway
Proximity	Far from user	Closer to user
Response Time	High	Low
Bandwidth Utilization	High	Low
Latency	High	Low
Network Congestion	High	Low
Resource Contention	High	Low
QoS	Low	High
Scalability	High	Medium
Distributed	No	Yes
Storage	Permanent	Temporary
Data Analytics	Slower	Faster

9.4 TREE-STRUCTURED TOPOLOGY (TST) FOR DISTRIBUTED AAAS

The hierarchical network architecture designed for using IoT devices, fog nodes, and the cloud is represented in Figure 9.2. The bottom layer, IoT devices, is responsible for sensing the data, and the data sensed is transferred to the next layer. The middle layer, the fog, is for distributed analytics of data. Either the task of analytics is completely done by a fog node at the middle layer or the fog in combination with the top layer, the cloud, together do analytics. A hierarchical network architecture can be represented using a tree, wherein a tree structure, there is a root node, a child node, and leaf nodes. In the same way, the proposed three-tier architecture consists of a root node as the cloud, its children as fog nodes, and sensors act as the children of the fog nodes. Also, sensors act as leaf nodes. A leaf node is the one that does not have any children. A tree-structured topology of height 2 is proposed for the provisioning of distributed analytics as a service. The cloud forms the highest level in the hierarchy, and sensors form the bottom level. When moving up in the hierarchy, the computational capacity and communication ability increase. At the middle layer, there can be any number of fog nodes, and in turn, each fog node can be associated with any number of sensors. The potential of storage, processing, and communication keeps on increasing while traversing from sensors, that is, leaf node, to the cloud (root node). The IoT devices keep on sensing the data and send the data to the fog node via branches, and the fog node does transmit data to the centralized cloud computing for later analytics.

The sensors that are present within the communication range of a fog node are termed as neighbors of the fog node. The mathematical expression for representing the neighborhood of nodes is given as a sensor, S_i, is termed as the neighborhood for the fog node, FOG_k, represented as N_{FOG_k}, if and only if it is present within the communication range of the fog node. Each fog node, FOG_k, can communicate with its siblings, FOG_j, which are at the same level. Also, fog nodes can interact with the cloud node, C_u, in the top layer. Each sensor, S_i, senses a particular feature, F_i, and thus, a fog node receives a contextual feature vector $\langle F_1, F_2, ... F_k \rangle$ from the sensors in the neighborhood of the fog node FOG_k at a discrete time interval $t \in \{1, 2, ..., T\}$. Also, each sensor S_i in N_{F_k} communicates with the fog node FOG_k at a regular time interval with transmission frequency f. However, the transmission frequency is influenced by the sampling rate and bandwidth. A fog node FOG_k will start doing analytics using the data gathered between the time interval t_1 and t_2. Based on the local storage capacity, the sensor can also store the transmitted feature in it for the specified duration D, and the new sensed features are concatenated at the end in the fog node. After the specified duration D, the transmitted features are discarded from the local storage of the particular sensor S_i.

9.5 DISTRIBUTED AAAS USING A FOG ENGINE

A fog engine that can communicate with IoT devices and the cloud through a network is proposed for distributed AaaS. Figure 9.3 shows the schematic diagram of the positioning of a fog engine to perform distributed analytics. A fog engine can perform data processing nearer to IoT devices, which results in a quick response time. A fog engine

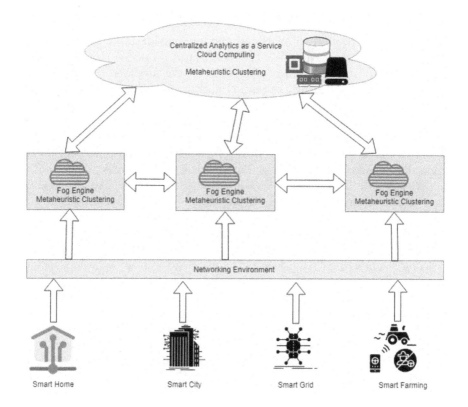

FIGURE 9.3 Distributed Analytics as a Service in Fog Computing.

can communicate with other fog engines, thereby creating a distributed environment for an efficient analysis of a user's request. A fog engine acts as a mediator that provides options for depositing the data to the cloud, which can then be used for later analytics. One of the functionalities of a mediator is that it made data from IoT devices reach the cloud, which is situated at remotely, although IoT devices are not directly connected to the cloud. Another functionality of a mediator is that it processes data for the sake of IoT devices. A fog engine is plugged with a metaheuristic clustering to perform optimal analytical requests from IoT devices. All fog engines are plugged with the same metaheuristic clustering, which is also available in the cloud for later analytics.

9.6 FFA-BASED FKM (FFAFKM) CLUSTERING FOR DISTRIBUTED ANALYTICS AS A SERVICE

The FFA is integrated with FKM to maximize the accuracy of FKMs for the optimal prediction of heart disease.

9.6.1 FFA FOR FINDING INITIAL CENTROID

The FFA is inspired by the characteristics of fireflies. The interaction between the firefly through the light flashed by them is the attractive feature that is used by Yang (43). When the distance increases, attractiveness decreases; that is, attractiveness is inversely proportional to distance. A firefly always gets attracted to a firefly with high brightness. Each firefly represents the initial centroid, and the fitness of the firefly represents the outcome of the FKM. Equation 9.1 represents the fitness function. The FFA intends to return the cluster centroid, which is taken as the initial centroid for FKM. The position of the firefly is computed as represented in Equation 9.2. The centroid returned by the FFA is taken by the FKM for finding optimal clusters. The centroid is then used to predict the class label of the new test instances.

$$Maximize\ Accuracy_{FKM} \qquad (9.1)$$

subject to

$$1\ MinimizeErrorRate_{FKM}$$

$$\left(Pos_{F_i}^d\right)_t = \left(Pos_{F_i}^d\right)_t + \beta_{dist_{F_iF_j}} * e^{-\gamma dist_{F_iF_j}^2} * \left[\left(Pos_{F_j}^d\right)_t - \left(Pos_{F_i}^d\right)_t\right] + \alpha_t \varepsilon_t, \qquad (9.2)$$

where γ determines the degree to which the updation of position depends on the distance between the fireflies, α represents the parameter for random movement of the fireflies, μ is a random value between 0 and 1, t represents iteration, and d represents dimension.

9.6.2 FKMs FOR THE OPTIMAL PREDICTION OF HEART DISEASE

FKM, a soft clustering method (44, 46, 48), finds the degree to which an instance belongs to a particular cluster. For the data set represented as $D \leftarrow \{x_1, x_2, ..., x_n\}$, the FKM partitions the data set into $|c|$ classes using the membership matrix U. The objective function for partitioning the instances in the data set is represented in Equation 9.3:

$$Minimize\ Z = \sum_{i=1}^{|D|}\sum_{j=1}^{|c|} u_{ij}^m d^2\left(x_i, c_j\right) \tag{9.3}$$

subject to

$$\sum_{a=1}^{|c|} u_{ab} = 1 \forall b \in \left[1, |D|\right], \tag{9.4}$$

where $|D|$ represents the number of instances in the data set,
$|c|$ represents the number of classes or cluster label,
$U = \left[u_{ij}\right]_{|D|*|C|}$ represents the degree of membership matrix, and
$d^2\left(x_i, c_j\right)$ represents the distance between the instance x_i and the cluster centroid c_j.

To minimize the objective function, the optimal cluster centroid is needed, which is found using the FFA. As a result, the clusters formed are optimal enough to predict the class label for the new instance. The optimal prediction of heart disease is essential as the significant insights can save human life. The algorithm for FFAFKM is shown in Algorithm 1.

Algorithm 1: $FFAFKM(\ )$
2   Input: Random population $P \leftarrow \{P_1, P_2, ..., P_n\}$; where each $P_i \in \{c_1, c_2\}$
3   Output: Class label for the new instances
4   While $T <= Max\,iteration$
5   For each Firefly FF_i
6   For each Firefly $F\,F_j$
7   Compute fitness $fit_T\left(FF_i\right)$ by invoking $FKM(\ )$
8   If $fit_T\left(FF_i\right) < fit_T\left(FF_j\right)$
9          $dist_{FF_iF_j} \leftarrow \sum_{d=1}^{dim\,\Sigma}\left|Pos_{FF_i}^d - Pos_{FF_j}^d\right|$
10         $\beta_{dist_{FF_iFF_j}} \leftarrow \beta_0 * e^{-\gamma\,dist_{FF_iFF_j}}$
11   Compute position of firefly as in Equation 2
12   Compute the new fitness $fit_T\left(FF_i\right)$
13   End If
14   End For
15   End For
16   End While
17         $cc \leftarrow \left(F_i\right)_{max\{f(F_i)\}}$
18   For each test instance T_i

19      $ClassLabel_{T_i} \leftarrow \left\{ c_i \mid \min\left(dist_{T_i c_i} \right) \right\}$

20   End For

9.7 PERFORMANCE EVALUATION

9.7.1 EXPERIMENTAL SETUP

The proposed metaheuristic clustering is tested with a real-world data set taken from the UCI repository. The experimentation is carried out in Java with 4 GB RAM and a 1.70-GHz processor. Actually, the experimentation is carried out in two different scenarios, one in which distributed data analytics is carried out by using two fog nodes that are created and the computation is carried out in fog node and the other in which the centralized analytics is performed in cloud computing. The details of the heart disease data set taken from the UCI repository (42) are given in Table 9.2. The experimentation is carried out using the model generated in ifogsim (49). The fog nodes and cloud nodes are integrated with the proposed FFAFKM. The configuration details of the nodes in the topology are represented in Table 9.3. Figures 9.4 and 9.5 represent the experimental setup and real-time setup in ifogsim. Figure 9.4 represents that each fog node is responsible to receive information from the sensors present in its area. The heart disease data set contains 13 attributes; thus, there are 13 sensors, whereby each sensor corresponds to the specified attribute. Each fog node is integrated with FFAFKM to predict the class label for each instance. Thus, the analytics is happened at the edge of the device, thereby minimizing latency. At the periodic interval, the

TABLE 9.2
Heart Disease Data Set Description

Attribute ID	Attribute Name
1	Age (in years)
2	Sex
3	Chest pain
4	Blood Pressure
5	Cholesterol
6	Sugar
7	Electrocardiographic Results
8	Heart Rate
9	Angina
10	Oldpeak
11	Slope
12	Number of Major Vessels Colored by Fluoroscopy
13	Type of Defect
14	Diagnosis of Heart Disease

TABLE 9.3

Experimental Configuration

Instance	RAM	Hard Disk	Power (Watts)	# instances
Sensors	2048 MB	200 KB	07.73	13
Fog Nodes	8	20	92.443	2
Cloud Node	32	150	117.45	1

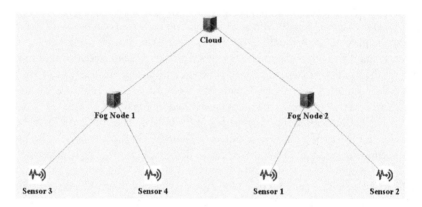

FIGURE 9.4 Design of WSNs through Fog Nodes.

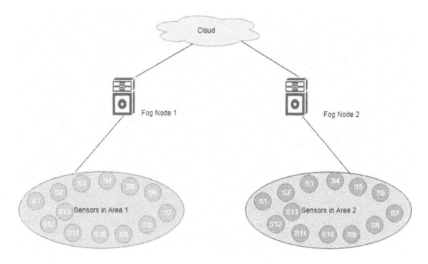

FIGURE 9.5 Real-Time Representation of the Proposed WSNs through Fog Computing.

fog nodes send the set of instances to the cloud, where FFAFKM is used for doing later analytics. A total number of 303 instances are present in the heart disease data set. The description of the attributes in the data set is depicted in Table 9.4.

TABLE 9.4
Description of Heart Disease Data Set

Attribute	Mean	Standard Deviation	Min	Max
Age (in years)	54.438944	9.038662	29	77
Sex	0.679868	0.467299	0	1
Chest pain	3.158416	0.60126	1	4
Resting Blood Pressure (in mm Hg)	131.689769	17.599748	94	200
Serum Cholesterol (in mg/dl)	246.693069	51.776918	126	564
Fasting Blood Sugar (in mg/dl)	0.148515	0.356198	0	1
Resting Electrocardiographic Results (in mV)	0.990099	0.994971	0	2
Maximum Heart Rate Achieved	149.607261	22.875003	71	202
Exercise-Induced Angina	0.326733	0.46794	0	1
Oldpeak	1.039604	1.161075	0	6.2
Slope	1.600660	0.616226	1	3
Number of Major Vessels Colored by Fluoroscopy	0.672241	0.931209	0	3
Type of Defect	4.734219	1.933272	3	7

9.7.2 EVALUATION OF THE FOG VERSUS THE CLOUD

9.7.2.1 Comparison of Accuracy, Root Mean Square Error, Sensitivity, Specificity, and False-Positive Rates

To show the efficiency of FFAFKM in fog computing, various investigations were conducted on the heart data set using different machine learning algorithms such as FKM. The efficiency of the proposed FFAFKM is good when compared to FKM, and the results obtained are shown in Table 9.5. Both algorithms attained almost the same level of accuracy, root mean square error, sensitivity, specificity, and false-positive rates in both the cloud and fog environments. However, when comparing the performance of the proposed FFAFKM, it achieves maximum level of accuracy, sensitivity, specificity, and minimum false-positive rates and the root mean square error. The reason behind the best performance of FFAFKM is that the metaheuristic algorithm FFA is used to compute optimal cluster centroid for the heart data set. In conventional FKM, random values are taken as centroid and thus the results are not optimal. The experimentation was carried out 20 times, and the mean value is considered for evaluating the performance of the algorithm in fog and cloud environments, and the outcomes are shown in Figure 9.6.

9.7.2.2 Comparison of Latency

Latency is defined as the propagation time that is the sum of time taken for transferring the request through the network to reach the intended destination and execution time of the request. Although a fog device does not have the rich computational and infrastructural capability, it resides closer to the sources that generate the data, and thus, the propagation time is less in the fog than in the cloud, where the resources responsible for

TABLE 9.5

Comparison of Performance of FFAFKM

Metrics	Fog AaaS		Cloud AaaS	
	FFAFKM	FKM	FFAFKM	FKM
Accuracy	0.9843	0.823	0.9843	0.823
Root Mean Square Error	0.1904	0.32	0.1904	0.32
Sensitivity	0.993	0.632	0.993	0.632
Specificity	0.992	0.813	0.992	0.813
False–Positive Rates	0.008	0.187	0.008	0.187

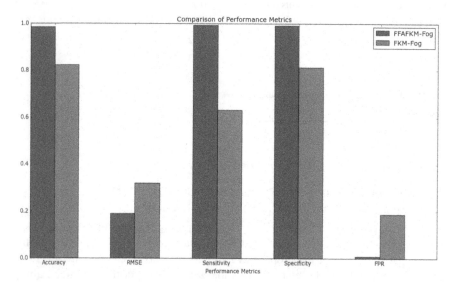

FIGURE 9.6 Comparison of Performance Metrics.

processing the user's request are present in a remote location, thereby incurring a significant delay in propagation time. Thus, the time required for data transfer is less in a fog environment, which has a significant reduction in delay of transferring the request. Also, if the size of the analytical request is not so big, then the execution time of the request does not have a huge impact regardless of whether the request is executed in the cloud or the fog. The experimentation is carried with the data request ranging from 50 to 250 patients, which is not so big, and therefore, the execution time is the same in the cloud and the fog, but as the cloud is too far from the data sources, latency is introduced for every batch of requests. As the fog is closer to the data sources, latency is minimal in a fog environment than in the cloud, as represented in Figure 9.7. The average minimal latency of the fog is 63.08% less than the cloud.

9.7.2.3 Comparison of Network Usage Time

Network usage time for fog and cloud environments is shown in Figure 9.8. From Figure 9.8, it is observed that the network usage time of the fog is less than the cloud

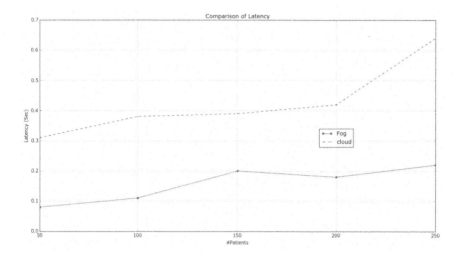

FIGURE 9.7 Comparison of Latency.

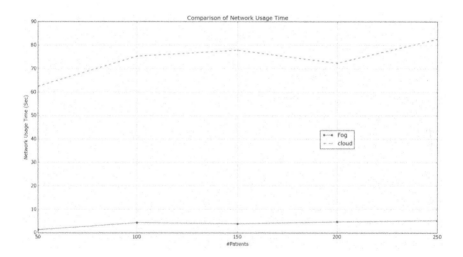

FIGURE 9.8 Comparison of Network Usage Time.

because of the ability of using local networking resources. Because the data sources that generate the data are located close to the fog nodes, the amount of network usage time is less in the fog, whereas in the case of the cloud environment, the time to transfer the data from fog to cloud is more as cloud nodes are situated at the remote. The experimentation is carried out by splitting the heart patients from the data set as 50 and keeps increasing by 50 for every time interval. It is seen that when the number of patients reaches 200, the network usage time for the cloud is decreased to 72.34 seconds from 77.8 seconds. Thus, fog analytics as a service achieves a huge benefit,

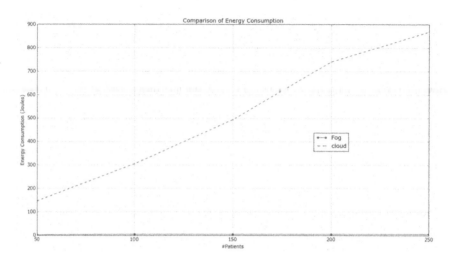

FIGURE 9.9   Comparison of Energy Consumption.

even though the network resources allocated are less. Also, from the experiments, it is revealed that the fog uses 94.76% fewer network resources than the cloud.

9.7.2.4 Comparison of Energy Consumption

From Figure 9.9, it is observed that the energy consumed by the fog nodes is much less when compared to the cloud nodes. The virtual machines consume more energy when performing the execution of an analytical request than do fog nodes. The experimentation is carried out by increasing the number of heart patients from 50 to 250, and in each case, the energy consumption is more for the cloud than the fog. The energy consumption of the fog when the number of heart patients is 250 is 75.5% less than the cloud. On average, the fog has a reduced energy consumption of 21.29% than the cloud.

9.8 CONCLUSION AND FUTURE WORK

Fog-based analytics as a service provisioning is introduced to minimize the delay, energy, and the usage of network that are very high in a traditional centralized cloud environment. Also, a new methodology based on a metaheuristic algorithm and a machine learning algorithm was proposed for the optimal provisioning of AaaS, and it was implemented for a heart disease data set. The proposed mechanism used a metaheursitic algorithm called FFA and integrated it with FKM to do optimal clustering. The centroid for the cluster is then used to forecast the classes of the instances in the heart data set, and the resultant accuracy, sensitivity, specificity, and false-positive rates were measured. Also, the experimental results show that FFAFKM achieves better results than other machine learning algorithms such as FKM.

REFERENCES

1. Mell, P., and T. Grance. *The NIST Definition of Cloud Computing*, https://nvlpubs.nist.gov/nistpubs/Legacy/SP/nistspecialpublication800-145.pdf.
2. Tordera, E. M. et al. (2016). What Is a Fog Node a Tutorial on Current Concepts Towards a Common Definition. *Arxiv Preprint*, arXiv:1611.09193, 28 November.
3. Yi, S., C. Li, and Q. Li. (2015). A Survey of Fog Computing: Concepts, Applications and Issues. *Proceedings of the 2015 Workshop on Mobile Big Data*, 37–42, 21 June.
4. Solutions, C. F. (2015). *Unleash the Power of the Internet of Things*. San Jose, CA: Cisco Systems Inc.
5. Kreutz, D. et al. (2015). Software-Defined Networking: A Comprehensive Survey. *Proceedings of the IEEE*, 103(1), 14–76.
6. Luan, T. H., L. Gao, Z. Li, Y. Xiang, G. Wei, and L. Sun. (2015). Fog Computing: Focusing on Mobile Users at the Edge. *Arxiv Preprint*, arXiv:1502.01815, 6 February.
7. Peng, M., S. Yan, K. Zhang, and C. Wang. (2015). Fog Computing Based Radio Access Networks: Issues and Challenges. *Arxiv Preprint*, arXiv:1506.04233, 13 June.
8. Dastjerdi, A. V., and R. Buyya. (2016). Fog Computing: Helping the Internet of Things Realize Its Potential. *Computer*, 49(8), 112–116, August.
9. Bonomi, F., R. Milito, P. Natarajan, and J. Zhu. (2014). Fog Computing: A Platform for Internet of Things and Analytics. In *Big Data and Internet of Things: A Roadmap for Smart Environments*. Cham: Springer, 169–186.
10. Yannuzzi, M., R. Milito, R. Serral-Gracià, D. Montero, and M. Nemirovsky. (2014). Key Ingredients in an IoT Recipe: Fog Computing, Cloud Computing, and More Fog Computing. *2014 IEEE 19th International Workshop on Computer Aided Modeling and Design of Communication Links and Networks (CAMAD)*, 325–329, 1 December.
11. Aazam, M., and E. N. Huh. (2014). Fog Computing and Smart Gateway Based Communication for Cloud of Things. *2014 International Conference on Future Internet of Things and Cloud*, 464–470, 27 August.
12. Bonomi, F., R. Milito, J. Zhu, and S. Addepalli. (2012). Fog Computing and Its Role in the Internet of Things. *Proceedings of the First Edition of the MCC Workshop on Mobile Cloud Computing*, 17, 13–16, 17 August.
13. Dubey, H. et al. (2015). Fog Data: Enhancing Telehealth Big Data Through Fog Computing. *Proceedings of the ASE Bigdata & Social Informatics*, 14, 7 October.
14. Barik, R. K. et al. (2016). FogGIS: Fog Computing for Geospatial Big Data Analytics. *2016 IEEE Uttar Pradesh Section International Conference on Electrical, Computer and Electronics Engineering (UPCON)*, 613–618, 9 December.
15. Arkian, H. R., A. Diyanat, and A. Pourkhalili. (2017). MIST: Fog-Based Data Analytics Scheme with Cost-Efficient Resource Provisioning for IoT Crowdsensing Applications. *Journal of Network and Computer Applications*, 82, 152–165, 15 March.
16. Al-Fuqaha, A., M. Guizani, M. Mohammadi, M. Aledhari, and M. Ayyash. (2015). Internet of Things: A Survey on Enabling Technologies, Protocols, and Applications. *IEEE Communications Surveys & Tutorials*, 17(4), 2347–2376, 15 June.
17. Jayaraman, P. P. et al. (2015). Scalable Energy-Efficient Distributed Data Analytics for Crowdsensing Applications in Mobile Environments. *IEEE Transactions on Computational Social Systems*, 2(3), 109–123, September.
18. Sharma, S. K., and X. Wang. (2017). Live Data Analytics with Collaborative Edge and Cloud Processing in Wireless IoT Networks. *IEEE Access*, 5, 4621–4635.
19. Alam, K. A., R. Ahmad, and K. Ko. (2017). Enabling Far-Edge Analytics: Performance Profiling of Frequent Pattern Mining Algorithms. *IEEE Access*, 5, 8236–8249.
20. Kulatunga, C., L. Shalloo, W. Donnelly, E. Robson, and S. Ivanov. (2017). Opportunistic Wireless Networking for Smart Dairy Farming. *IT Professional*, 19(2), 16–23.

21. Kaur, K., T. Dhand, N. Kumar, and S. Zeadally. (2017). Container-as-a-Service at the Edge: Trade-Off Between Energy Efficiency and Service Availability at Fog Nano Data Centers. *IEEE Wireless Communications*, 24(3), 48–56, June.

22. He, J., J. Wei, K. Chen, Z. Tang, Y. Zhou, and Y. Zhang. (2018). Multitier Fog Computing with Large-Scale IoT Data Analytics for Smart Cities. *IEEE Internet of Things Journal*, 5(2), 677–686, April.

23. Yang, S. (2017). IoT Stream Processing and Analytics in the Fog. *IEEE Communications Magazine*, 55(8), 21–27.

24. Raafat, H. M. et al. (2017). Fog Intelligence for Real-Time IoT Sensor Data Analytics. *IEEE Access*, 5, 24062–24069.

25. Patel, P., M. I. Ali, and A. Sheth. (2017). On Using the Intelligent Edge for IoT Analytics. *IEEE Intelligent Systems*, 32(5), 64–69, September.

26. Chaudhary, R., N. Kumar, and S. Zeadally. (2017). Network Service Chaining in Fog and Cloud Computing for the 5G Environment: Data Management and Security Challenges. *IEEE Communications Magazine*, 55(11), 114–122, November.

27. El-Sayed, H., S. Sankar, M. Prasad, D. Puthal, A. Gupta, M. Mohanty, and C. T. Lin. (2018). Edge of Things: The Big Picture on the Integration of Edge, IoT and the Cloud in a Distributed Computing Environment. *IEEE Access*, 6, 1706–1717.

28. Darwish, T. S., and K. A. Bakar. (2018). Fog Based Intelligent Transportation Big Data Analytics in the Internet of Vehicles Environment: Motivations, Architecture, Challenges, and Critical Issues. *IEEE Access*, 6, 15679–15701.

29. Abdelwahab, S., S. Zhang, A. Greenacre, K. Ovesen, K. Bergman, and B. Hamdaoui. (2018). When Clones Flock Near the Fog. *IEEE Internet of Things Journal*, 5(3), 1914–1923, June.

30. Xu, X., L. Zhang, S. Sotiriadis, E. Asimakopoulou, M. Li, and N. Bessis. (2018). CLOTHO: A Large-Scale Internet of Things-Based Crowd Evacuation Planning System for Disaster Management. *IEEE Internet of Things Journal*, 5(5), 3559–3568, October.

31. Diro, A. A., N. Chilamkurti, and Y. Nam. (2018). Analysis of Lightweight Encryption Scheme for Fog-to-Things Communication. *IEEE Access*, 6, 26820–26830.

32. Liu, G., S. Liu, K. Muhammad, A. K. Sangaiah, and F. Doctor. (2018). Object Tracking in Vary Lighting Conditions for Fog Based Intelligent Surveillance of Public Spaces. *IEEE Access*, 6, 29283–29296.

33. Yacchirema, D. C., D. Sarabia-Jácome, C. E. Palau, and M. Esteve. (2018). A Smart System for Sleep Monitoring by Integrating IoT with Big Data Analytics. *IEEE Access*, 6, 35988–6001.

34. Sharma, P. K., S. Rathore, Y. S. Jeong, and J. H. Park. (2018). SoftEdgeNet: SDN Based Energy-Efficient Distributed Network Architecture for Edge Computing. *IEEE Communications Magazine*, 56(12), 104–111, December.

35. Iqbal, R., T. A. Butt, M. O. Shafique, M. W. Talib, and T. Umer. (2018). Context-Aware Data-Driven Intelligent Framework for Fog Infrastructures in Internet of Vehicles. *IEEE Access*, 6, 58182–58194.

36. Wang, Y., M. C. Meyer, and J. Wang. (2019). Real-Time Delay Minimization for Data Processing in Wirelessly Networked Disaster Areas. *IEEE Access*, 7, 2928–2937.

37. Taneja, M., N. Jalodia, and A. Davy. (2019). Distributed Decomposed Data Analytics in Fog Enabled IoT Deployments. *IEEE Access*, 7, 40969–40981, 27 March.

38. Pabitha, R. (2017). Novel Car Parking System Using Swarm Intelligence. *International Journal of Pure and Applied Mathematics*, 117(16), 289–298.

39. Keerthana, R., C. Haripriya, P. Pabitha, and R. S. Moorthy. (2018). An Efficient Cancer Prediction Mechanism Using SA and SVM. *2018 Second International Conference on Intelligent Computing and Control Systems (ICICCS)*, 1140–1145, 14 June.

40. Ramakrishnan, R., M. S. Ram, P. Pabitha, and R. S. Moorthy. (2018). Freezing of Gait Prediction in Parkinsons Patients Using Neural Network. *2018 Second International Conference on Intelligent Computing and Control Systems (ICICCS)*, 61–66, 14 June.

41. Scirè, A., F. Tropeano, A. Anagnostopoulos, and I. Chatzigiannakis. (2019). Fog-Computing-Based Heartbeat Detection and Arrhythmia Classification Using Machine Learning. *Algorithms*, 12(2), 32, February.

42. Janosi, Andras, William Steinbrunn, Matthias Pfisterer, and Robert Detrano. *Hear Disease Dataset: UCI Repository*, https://data.world/uci/heart-disease.

43. Yang, X. S. (2010). Firefly Algorithm, Levy Flights and Global Optimization. In *Research and Development in Intelligent Systems*. London: Springer, vol. XXVI, 209–218.

44. Jain, A. K. (2010). Data Clustering: 50 Years Beyond k-Means. *Pattern Recognition Letters*, 31(8), 651–666, 1 July.

45. Parameswaran, P., and R. Shenbaga Moorthy. (2019). Secure Pervasive Healthcare System and Diabetes Prediction Using Heuristic Algorithm. *Intelligent Pervasive Computing Systems for Smarter Healthcare*, 179–205, 22 July.

46. Dunn, J. C. (1974). Well-Separated Clusters and Optimal Fuzzy Partitions. *Journal of Cybernetics*, 4(1), 95–104, 1 January.

47. Shenbaga Moorthy, R., and P. Pabitha. (2019). Optimal Provisioning and Scheduling of Analytics as a Service in Cloud Computing. *Transactions on Emerging Telecommunications Technologies*, e3609.

48. Huang, Z., and M. K. Ng. (1999). A Fuzzy k-Modes Algorithm for Clustering Categorical Data. *IEEE Transactions on Fuzzy Systems*, 7(4), 446–452, August.

49. Gupta, H., A. Vahid Dastjerdi, S. K. Ghosh, and R. Buyya. (2017). iFogSim: A Toolkit for Modeling and Simulation of Resource Management Techniques in the Internet of Things, Edge and Fog Computing Environments. *Software: Practice and Experience*, 47(9), 1275–1296, September.

50. https://blogs.cisco.com/digital/fog-analytics-turning-data-into-real-time-insight-and-action.

51. Shenbaga Moorthy, R., and P. Pabitha. (2019). Optimal Provisioning and Scheduling of Analytics as a Service in Cloud Computing. *Transactions on Emerging Telecommunications Technologies*, 9, e3609, 30 September.

10 DLA-RL
Distributed Link Aware-Reinforcement Learning Algorithm for Delay-Sensitive Networks

O.S. Gnana Prakasi, P. Kanmani, K. Suganthi,
G. Rajesh, and T. Samraj Lawrence

CONTENTS

10.1 INTRODUCTION

The exponential growth of the internet has led to the transmission of a huge volume of delay-sensitive data between the mobile nodes in a very dynamic delay-tolerant environment. This huge volume of delay-sensitive data includes data from health care, which helps monitor the patients remotely; online transactions, which require secure communication between the nodes; live webinars for the learning process and involves a large number of users sharing the information without any network failure, from the information technology industry, entertainment movies, and the like and are transmitted across the nodes. These result in a drastic increase in the data traffic globally in very dynamic ad hoc networks where the routes are not stable.

A mobile ad hoc network is an example of a dynamic network in which nodes move from one position to another at different speeds at different times. Because the nodes are moving continuously, the links between these nodes break, resulting in path failure, and the packets in these paths cannot reach their destinations on time. The delay is due to reestablishing an alternate path from the available nodes. This reduces the performance of the network, especially in the delay in packet delivery, which makes the data unuseful at the destination. Again, routing overhead is also increased due to frequent path disconnections and data replication flooding the entire work when finding alternate paths. Routing these delay-sensitive data from the source to the destination in such a dynamic environment is a challenging task. So it is necessary to find an optimal path with a stable and reliable link between the source and the destination.

Incorporating machine learning algorithms (4) can provide opportunistic and nondeterministic in a dynamic delay-tolerant network. Thus, machine learning-based routing algorithms provide benefits by learning from the past to predict the future. Reinforcement learning (RL) is one of the most commonly used machine learning algorithms, in which the learner discovers how to achieve a desired outcome by maximizing a numerical reward (15, 19).

Unlike supervised learning, RL maintains the best balance between exploration and exploitation of a dynamic network (17, 18). Focusing on link stability between the nodes and the speed of each node in the network improves the usability of the dynamic network. Temporal difference (TD) learning is one of the most common RL methods. This approach computes the difference between the expected outcome and the actual outcome. The learning helps with reducing the deviation between the actual outcome and the expected outcome by setting appropriate parameters. The main advantage of TD is that it does not require any training data set to train the entire system using machine learning algorithms. Instead, the agent uses results generated by the sequence of previous states for predicting the expected outcome of the current variable (16). Applying TD learning in establishing routing paths between the source and the destination enables remote monitoring, and hence, an optimal path with a high-reliability link can be established between the source and the destination node.

Thus, in this chapter, we propose a DLA-RL: Distributed Link Aware-Reinforcement Learning algorithm for delay-sensitive networks, which uses Boltzmann exploration (18) for learning the link using a link metric (LM) between any two mobile nodes. Thus, the reliable path with minimal delay is selected between the source and the destination by adopting the temporal function of a reinforcement algorithm. Hence, the performance of the proposed work is increased compared to traditional routing algorithms.

10.2 LITERATURE SURVEY

Hao et al. (20) constructed a learning automata theory–based feedback mechanism to optimize router selection from the available routes using a convergence algorithm for the MANET environment. Gnana and Varalakshmi (26) introduced a decision-making process to select the optimal path is executed through a decision tree using Decision Tree-Based Routing Protocol (DTRP). Hence, an optimal path can be selected from the node properties. Sakthivel et al. (22) suggested on-demand Progressive Distance Vector (PAODV) extends the routing process by switching the traffic to the optimal path and shrinks the current transmission path. Khan et al. (23), using Location-Aware Grid-Based Hierarchical Routing (LGHR), constructed a grid in non-overlapping zones to identify a stable path in a dynamic environment in which nodes were moving at high velocities. Gnanaprakasi and Varalakshmi (10) used a fuzzy approach to construct an optimal path by considering routing parameters. Haripriya et al. (24) optimized the flooding mechanism by handling packet duplication and errors generated as a result of the noise in a wireless environment with help of a dynamic probabilistic broadcasting algorithm. Saravanan (25) focused on energy conversation and quality of service by choosing a path that consumes minimal energy during its transmission of packets.

Tao Wang and Hung and Suganthi et al. (1, 21) proposed probabilistic analysis to find the most reliable route by creating reliable clusters based on the node properties. (2) introduced interface direction in probabilistic routing known as Directional Routing Table (DRT), which maps the set of IDs to each interface of the mobile nodes. Xiufeng Wang et al. (3) show how to estimate the durability of the connection to another vehicle using an extended link duration prediction (ELDP) model. Sarao (5) analyzed the issues and challenges while applying deep learning and machine learning algorithms in a wireless environment and how to predict and recover from abnormalities in a wireless environment. Hernández-Jiménez et al. (6) proposed a deep learning–based router in which the best messages are selected by the scheduler to transmit to the best hop in the path to reach the destination. Dudukovich et al. (7) have shown that RL with a Bayesian learning process helps in making the routing decisions for selecting the optimal path in delay-tolerant networks. Docition and Varghese (8) proposed how to learn data flow rates using an RL approach for selecting the appropriate neighbors to forward the packets. Hence, the quality of service can be improved in a wireless network. Liu et al. (9) proposed a distributed energy-efficient RL for routing in wide-area wireless mesh IoT networks by analyzing the failure rate.

Boushaba et al. (11) used the routing metric reinforcement learning-based best path routing to balance the load among the nodes using RL during transmission. Mitchel et al. (12) suggested how the available spectrum can be used by reserving the spectrum and pre-partitioning the spectrum using an RL approach in cognitive radio networks. Bennis et al. (13) established a decentralized and self-configuring approach that uses the local knowledge of the nodes to apply an RL strategy to predict the routing path. Ho (14) introduced adaptive exploration techniques as a learning process to identify the possible routes for wireless networks.

From the survey, it can be suggested that applying an RL algorithm in routing helps identify an optimal path between the source and the destination by learning the previous paths of transmission. Hence, we proposed a new DLA-RL, which uses an LM and the node speed as parameters for learning the routing strategy between the source and the destination. Thus, the selected optimal path minimizes the delay by minimizing the average link failure.

10.3 DLA-RL ALGORITHM

10.3.1 System Model

Transferring packets from the source node (SN) to a destination node (DN) through one or more intermediate nodes (INs) in a dynamic delay-sensitive environment is a very challenging task. The system model in this chapter is built with a very dynamic set of mobile nodes that are moving randomly at varying speeds at different times. Each node is allowed to communicate with its neighbor only if they are in communication range and can ensure a reliable path with a sufficient link quality. The nodes are moving randomly at varying speeds, making the routing path very dynamic.

Learning the routing path by using an RL algorithm helps identify and adopt changes in the dynamic network, and hence, we can find a reliable path for enabling communication. Link reliability helps analyze the stability of the link, and the relative speed of mobile speaks to its dynamic nature. Learning these properties with the proposed DLA-RL algorithm helps find the optimal path with minimal delay and minimal link failure. Thus, the delay in packet transfer can be minimized by reducing the route reestablishment.

If a mobile node wants to communicate with any other node, it checks for the next hop (NH) in the routing table from the list of one-hop neighbor and selects the node with a higher probability of delivering the packets to their destination. The routing decision process and the functionalities of each mobile node in finding the optimal route using the proposed DLA-RL algorithm are illustrated in Figure 10.1. The process is initiated by the SN, which is ready for transmission. In this scenario, each

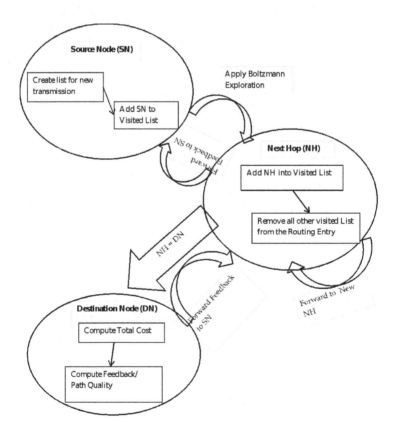

FIGURE 10.1   Routing Process in DLA-RL.

mobile node is allowed to learn the network individually and makes the local decision which can be distributed within the entire network. First, the SN selects the NH from its routing table by applying the Boltzmann exploration and adds the node to a visited list, which is updated in the packet header. The NH, on receiving the packet, checks the destination and updates, with the feedback, the SN in terms of path quality, which is computed in Equation 10.5. The path quality is forwarded to the SN through the intermediate nodes in the reverse path. If the current node is an intermediate node, it removes all previously visited nodes before forwarding the packet to the NH. Thus, the proposed DLA-RL selects the path in the dynamic delay-sensitive environment by learning node properties and the link properties.

To learn, in this model, we introduce an additional LM, which helps with the probability of selecting a particular node as the NH from the list of one-hop neighbors. This probability of selecting the NH is calculated using the Boltzmann exploration, and it is given in the Equation 10.1: (18)

$$P\left(NH_n\right) = \frac{e^{\frac{LM(NH_n)}{\tau}}}{\sum_{i=1}^{n} e^{\frac{LM(NH_n)}{\tau}}}. \tag{10.1}$$

$P\left(NH_n\right)$ is defined as the probability of selecting the nth node as the NH. τ is the control of the spread of a softmax distribution of all possible routes. LM generates the probability using the Boltzmann process, which helps RL adopt changes and optimally evaluate the best path.

If a source node is ready to transmit the data to the destination, a new transmission entry is initiated in routing table for that DN. Initially, the LM is computed as an average value of its entire one-hop neighbor as the path is still not established. But for an existing path, it is computed using the Boltzmann exploration. Once the packet is dispatched from the source node to its next-hop, the source node waits for its acknowledgment as the feedback/path quality from the destination node.

Now, once the SN receives the feedback from the destination, it updates the LM of all visited nodes using the received link reliability, the speed of the nodes, and the path reliability using RL.

The cost of the transmission is defined as a cost involved in establishing the point-to-point link between the SN to an IN or any IN to the next IN or the last IN to the DN. When moving from the SN to the DN, every IN will be added as visited nodes in the header, and this visited node will be removed from the list if that node is selected as the NH. A transmission is known to be unsuccessful if it cannot reach the destination through INs.

Now we compute the cost function by considering link reliability along with the relative speed of the nodes, which helps the learning algorithm to adapt the changes by the nodes during the transmission in a delay-sensitive dynamic environment. To update this, first we compute the cost of the current transmission between any two INs using Equation 10.2:

$$C = W_1 R(l_x) - (W_2 loglog\, S_{TN}) - (W_3 loglog\, S_{RN}), \tag{10.2}$$

where W_1, W_2, W_3 are some randomly assigned prescribed positive weights for learning the system; S_{TN} and S_{RN} are the speeds of the current transmitting and the receiving nodes, respectively; and $R(l_x)$ is defined as the link reliability, and it is computed using Equation 10.3:

$$R(l_x) = \{\int_t^t T_x(T) dT \, T_x > 0 \, 0 \, otherwise, \tag{10.3}$$

where $T_x(T)$ is defined as the probability density function of the communication duration T between the transmitting node and the receiving node during the interval t and is computed from distance between the nodes, its communication range, and the relative velocity of each node.

At the other end, once the transmission is successful, now we calculate the total cost to transmit a packet from the SN to the DN, and it given by Equation 10.4:

$$C_{total} = min \ (R(l_{x1}), R(l_{x2}), \ldots \ldots, R(l_{xn})), \tag{10.4}$$

where $R(l_{x1})$ is the link reliability of the first link (from the SN to the first IN) and $R(l_{xn})$ is the link reliability from the last IN and the DN. Thus, the path quality (PQ) is defined in terms of route reliability and the success bonus (SB). This value will be passed to the RL algorithm and will update the same in the routing table of each node after retrieving the feedback from the destination node to the source node. The formulas for the PQ and the SB are shown in Equations 10.5 and 10.6:

$$PQ = SB - C_{total} \tag{10.5}$$

$$SB = \{SBV \quad , if \ transmission \ is \ successful$$
$$0 \quad , if \ the \ transmission \ is \ unsucessful. \tag{10.6}$$

Equation 10.5 calculates the PQ as the total cost to reach the destination route cost C_{total} as well as the SB. If the transmission is successful a configurable predefined positive SB value (SBV) will be added to a successful transmission to compensate the total cost values. Zero SB will be given to unsuccessful transmissions, and in this case, PQ will be negative to discourage routing using the same path in the future.

10.3.2 TD Learning

In this chapter, we invoke a TD-based learning approach to update the LM in the routing table of each mobile node once the packet has reached the destination and the feedback from the DN has been sent. Once the feedback is received, the Boltzmann exploration process is used to balance the exploration using TD learning when selecting the NH from the list of one-hop neighbors. Because TD uses only recent transmissions and its feedback for predicting the next value, the computational overhead and storage are minimized. In addition to the LM, we add the time the node visited the variable in the routing table to ensure the time the route has been selected and all the possible NHs to reach the destination are visited before a new LM value is computed. Thus, the new LM is computed as given in Equation 10.7 (9):

$$LM_{new}(NH) = LM(NH) + \beta(PQ + \gamma LM(NH') - LM(NH)). \tag{10.7}$$

The updated new LM ($LM_{new}(NH)$) is computed from the current LM value at the node LM(NH), the PQ, and the expected outcome ($LM(NH')$), which is computed as the average of all LMs of all possible NHs in the current transmission. In addition, two control parameters, β and γ, denote the learning rate and the discount rate, respectively, showing how the algorithm learns the current network and predicts the optimal path for the data transmission. Figure 10.2 shows the computing procedure of the LM and the updating of the routing table using Boltzmann exploration.

Three parameters are used as part of the learning process; among these, two parameters are used in temporal learning to determine the outcome of the process. One more parameter, τ, is used in the decision-making to select the optimal path between the source and the destination and explore the possibility to use all the routes in the network. All these parameters have an impact on building the network and maintaining the route.

```
//Algorithm: Computing Link Metric (LM)
//Input: Nodes, node speed, τ, β, γ
//Output: Link Metric, Path Quality and routing path.
Initializes the variable
LN=Average of LN of its entire one-hop neighbour
//Selects Next-Hop
While (NH!=DN) do
        //selects Next-Hop
        // Compute the Cost
        If
                R(lₓ)
        Else
                R(lₓ)
        End while
If (SN==DN)
        //update LM from PQ
        C_total = min ()
        If (transmission is success)
                SB= SBV
        PQ = SB—C_total
Update PQ to source node.
//Update to new LM
LM_new(NH) = LM(NH) + β(PQ + γLM(NH') – LM(NH))
```

FIGURE 10.2 Algorithm to Compute LM.

10.3.2.1 Hyper-Parameter, τ

The process of balancing, exploring, and exploiting are given the Boltzmann exploration. Parameter τ controls the output by converting LM values into the probability of selecting the link. In addition, a softmax function is used to normalize the values of the output. When $(\tau \to \infty)$, the possibility is to select all the nodes and invoke a random process to select the NH. But when $(\tau \to 0)$, the probability reflects the real difference between the possible selection. Thus, by controlling the value of τ, one can balance exploring new and low LM values and hence involve them in selecting the optimal path.

10.3.2.2 Discount Rate, γ

The discount rate is defined in the range $\in (0, 1)$, which is multiplied by the expected LM value. Balancing the discount rate γ has a great impact on learning the network, and the performance of the network is increased by selecting the optimal path. If γ is greater than 1, a large value will be added, and hence, the output value gets diverged. If the value is too small, future impacts may be ignored while learning the network.

10.3.2.3 Learning Rate, β

The learning rate, β, is also known as a step-size process that is multiplied by the new information calculated from the new PQ value as well as the TD. When β is set to zero, the learning is zero and hence ignores the new learnt information. An optimal β value leads in the best learning of the network so that errors can be minimized.

10.4 SIMULATION AND RESULTS

The simulation is carried out using a python simulator; the simulation environment consists of a set of mobile nodes that carry delay-sensitive information from the source to the destination. These nodes are allowed to move at different speeds at different times. All the nodes are allowed to move independently in the given region. The results are analyzed with nodes 25, 50, and 100 with a simulation time of 100 seconds. The simulations performed with different Boltzmann exploration parameter (τ), discount rate (γ), and learning rate (β) with five different rates such as 0.1, 0.2, 0.5, 0.8 and 1. The graph in Figure 10.3 shows the average link failure rate with various learning parameters that are applied to the DLA-RL algorithm in a mobile environment. The nodes are allowed to generate the packets randomly during the simulation, and the performance is analyzed in terms of throughput, delay, spectral efficiency and failure rates with different hyper-parameters, discount rate and learning rate. These performances are compared and analyzed with an existing DTRP and centralized Shortest Path First (SPF) routing algorithms. The routing strategy of the DTRP (25) is based on a decision tree in which the learning with the feedback is not applicable, whereas the centralized SPF (9) routing strategy is based on the reinforcement algorithm, but mobility is not considered. Hence, the performance is degraded in the dynamic environment.

Throughput is measured as the transmission rate of the mobile nodes. Figure 10.4 shows the throughput of different routing algorithms with different

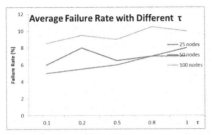
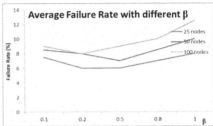

FIGURE 10.3a   Failure rate with different τ   **FIGURE 10.3b**   Failure rate with different β

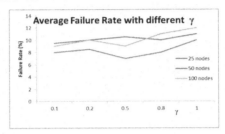

FIGURE 10.3c   Failure rate with different γ

FIGURE 10.3   Average Failure Rate.

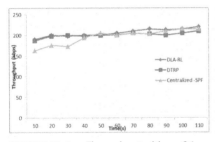

FIGURE 10.4a   Throughput with τ = 0.1, β = 0.1, ν = 0.1

FIGURE 10.4b   Throughput with τ = 0.5, β = 0.5, ν = 0.5

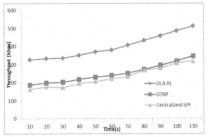
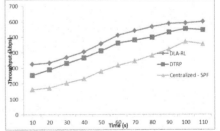

FIGURE 10.4c   Throughput with τ = 0.8, β = 0.8, ν = 0.8

FIGURE 10.4d   Throughput with τ = 0.5, β = 0.8, ν = 0.5

FIGURE 10.4   Throughput Analysis.

learning parameters. Although the figure shows no improvement with rates 0.1, there is an improvement in the performance of the DLA-RL, with higher learning rates compared to the DTRP algorithm and the centralized SPF routing algorithm.

Figure 10.5 shows the end-to-end delays of the routing algorithms, and it is compared to other routing algorithms. Because the data are delay-sensitive, the end-to-end delay needs to be very minimal. The figure shows that the end-to-end delay in centralized SPF is very high compared to the DLA-RL and the DTRP due to mobility. Again, the figure shows that the DLA-RL outperforms by minimizing the delay drastically by learning the dynamic environment more accurately and feeding the same learning to find the optimal path for packet transmission. Hence, the delay is minimized by the proposed DLA-RL routing algorithm.

The failure rate is measured as the average number of link failures between any two nodes. As nodes are moving at different speeds in different directions, the possibility for the node to move out of the transmission rate will also increase. So we have to select the link with the higher reliability value. By learning the link properties with link reliability and the relative speed of the nodes, the optimal link is selected. Figure 10.6 shows that the failure rate in DLA-RL is drastically reduced compared to the DTRP and centralized SPF.

Spectral efficiency is defined in terms of data rate and the bandwidth that states the payload and its overhead. Figure 10.7 shows the comparison of spectral efficiency

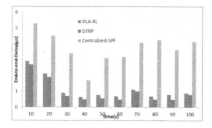

FIGURE 10.5a   Average Delay $\tau = 0.1$, $\beta = 0.1$,

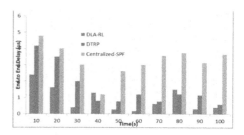

FIGURE 10.5b   Average Delay $\tau = 0.5$, $\beta = 0.5$, $\gamma = 0.5$

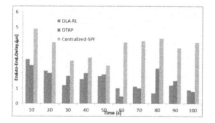

FIGURE 10.5c   Average Delay $\tau = 0.8$, $\beta = 0.8$, $\gamma = 0.8$

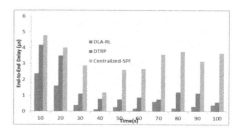

FIGURE 10.5b   Average Delay $\tau = 0.5$, $\beta = 0.8$, $\gamma = 0.5$

FIGURE 10.5   Average End-to-End Delay.

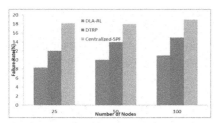

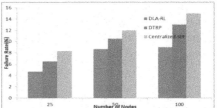

FIGURE 10.6a   Failure Rate $\tau = 0.1$, $\beta = 0.1$, $\gamma = 0.1$ **FIGURE 10.6b**   Failure Rate $\tau = 0.5$, $\beta = 0.5$, γ

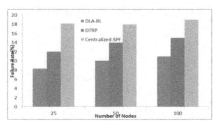

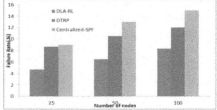

FIGURE 10.6a   Failure Rate $\tau = 0.8$, $\beta = 0.8$, $\gamma = 0.8$ **FIGURE 10.6d**   Failure Rate $\tau = 0.5$, $\beta = 0.8$, $\gamma = 0.5$

FIGURE 10.6   Average Failure Rate.

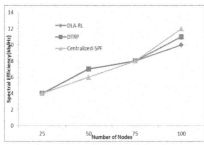

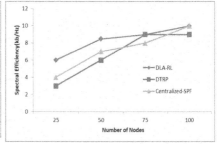

FIGURE 10.7a   Spectral Efficiency $\tau = 0.1$,   **FIGURE 10.7b**   Spectral Efficiency $\tau = 0.5$,
$\beta = 0.1$, $\gamma = 0.1$ $\beta = 0.5$, $\gamma = 0.5$

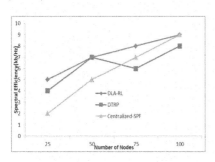

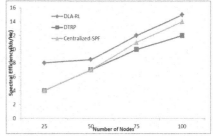

FIGURE 10.7c   Spectral Efficiency $\tau = 0.8$,   **FIGURE 10.7d**   Spectral Efficiency $\tau = 0.5$,
$\beta = 0.8$, $\gamma = 0.8$ $\beta = 0.8$, $\gamma = 0.5$

FIGURE 10.7   Spectral Efficiency.

with other routing algorithms. The figure shows little improvement in the spectral efficiency with other algorithms; we find improvement in Figure 10.7d with the learning parameters $\tau = 0.5$, $\beta = 0.8$, $\gamma = 0.5$. The main focus of this algorithm is to find the optimal path selection but not with spectral usage. In the future, this can be improved by considering other spectral parameters.

10.5 CONCLUSION

Using an RL algorithm in a dynamic delay-sensitive environment helps with learning the network by learning the node and the link properties in the routing path to select the optimal path with minimal loss. DLA-RL analyzes the routing with the parameters' link reliability and the nodes' speed to compute the LM, which selects the NH using the Boltzmann exploration and updating the PQ of the route. Thus, this algorithm helps keep track of the path, and hence, the average lifetime of the route is increased and the average end-to-end delay is minimized. The experiment results show that the performance enhancement of the delay-sensitive network is improved when compared to other routing algorithms.

10.6 GLOSSARY

Link Reliability: Predicts the future availability of the link using the active link between any two nodes.
Path Quality: The feedback calculated as a successful transmission.
Temporal Learning: Computes the difference between the expected outcome and actual outcome from the previous transmission.

REFERENCES

1. Wang, Tao, and William N. N. Hung. (2013). Reliable Node Clustering for Mobile Ad Hoc Networks. *Journal of Applied Mathematics*, 8.
2. Cheng, B., M. Yuksel, and S. Kalyanaraman. (2010). Using Directionality in Mobile Routing. *Wireless Networks*, 16, 2065–2086.
3. Wang, Xiufeng, Chunmeng Wang, Gang Cui, Qing Yang, and Xuehai Zhang. (2016). ELDP: Extended Link Duration Prediction Model for Vehicular Networks. *International Journal of Distributed Sensor Networks*, 12(4), 1–21.
4. Zantalis, Fotios, Grigorios Koulouras, Sotiris Karabetsos, and Dionisis Kandris. (2019). A Review of Machine Learning and IoT in Smart Transportation. *Future Internet*, 11(94), 1–23.
5. Sarao, Pushpender. (2019). Machine Learning and Deep Learning Techniques on WirelessNetworks. *International Journal of Engineering Research and Technology*, 12(3), 311–320.
6. Hernández-Jiménez, Roberto, Cesar Cardenas, and David Muñoz Rodríguez. (2019). Modeling and Solution of the Routing Problem in Vehicular Delay-Tolerant Networks: A Dual, Deep Learning Perspective. *Applied Science*, 9(5254), 1–17.
7. Dudukovich, Rachel, Alan Hylton, and Christos Papachristou. (2017). A Machine Learning Concept for DTN Routing. *IEEE International Conference on Wireless for Space and Extreme Environments (WiSEE)*, 110–115.

8.   Docition Simi, S., and Sruthi Ann Varghese. (2015). Enhance QoS by Learning Data Flow Rates in Wireless Networks Using Hierarchical. *4th International Conference on Eco-friendly Computing and Communication Systems, ICECCS, Procedia Computer Science*, 70, 708–714.

9.   Liu, Y., K. Tong, and K. Wong. (2019). Reinforcement Learning Based Routing for Energy Sensitive Wireless Mesh IoT Networks. *Electronics Letters*, 55(17), 966–968.

10. Gnanaprakasi, O. S., and P. Varalakshmi. (2016). EFG-AOMDV: Evolving Fuzzy Based Graph—AOMDV Protocol for Reliable and Stable Routing in Mobile Ad hoc Networks. *Ad Hoc & Sensor Wireless Networks*, 33, 1–24.

11. Boushaba, M., A. Hafid, and A. Belbekkouche. (2011). Reinforcement Learning Based Best Path to Best Gateway Scheme for Wireless Mesh Networks. *IEEE 7th International Conference on Wireless and Mobile Computing, Networking and Communications (WiMob)*, https://www.researchgate.net/publication/221508245_Reinforcement_learning-based_best_path_to_best_gateway_scheme_for_wireless_mesh_networks.

12. Mitchell, P., T. Jiang, and D. Grace. (2011). Efficient Exploration in Reinforcement Learning-Based Cognitive Radio Spectrum Sharing. *IET Communications*, 5(10), 1309–1317.

13. Bennis, M., S. Perlaza, P. M. Blasco, Z. Han, and H. V. Poor. (2013). Self Organization in Small Cell Networks: A Reinforcement Learning Approach. *IEEE Transactions on Wireless Communications*, 12(7), 3202–3212.

14. Ho Chi Minh City. (2017). Adaptive Exploration Strategies for Reinforcement Learning. *International Conference on System Science and Engineering*, https://ieeexplore.ieee.org/document/8030828.

15. Barto, A., and R. Sutton. (1998). *Reinforcement Learning: An Introduction*, Thomas Dietterich, Ed. Cambridge: MIT Press.

16. McClelland, J. L. (2015). Temporal-Difference Learning. *Explorations in Parallel Distributed Processing: A Handbook of Models, Programs, and Exercises*. San Jose: Standford University, 193–216.

17. Tang, F., B. Mao, Z. M. Fadlullah, N. Kato, O. Akashi, T. Inoue, K. Mizutani. (2018). On Removing Routing Protocol from Future Wireless Networks: A Real-time Deep Learning Approach for Intelligent Traffic Control. *IEEE Wireless Communications*, 25(1), 154–160.

18. Cesa-Bianchi, N., C. Gentile, G. Lugosi, and G. Neu. (2017). Boltzmann Exploration Done Right. In *31st Conference on Neural Information Processing Systems*. Long Beach, CA: NIPS.

19. Kaelbling, L. P., M. L. Littman, and A. W. Moore. (1996). Reinforcement Learning: A Survey. *Journal of Artificial Intelligence Research*, 4, 237–285.

20. Hao, S., H. Zhang, and M. Song. (2018). A Stable and Energy-Efficient Routing Algorithm Based on Learning Automata Theory for MANET. *Journal of Communications and Information Networks*, 3, 52–66.

21. Suganthi, K., B. Vinayagasundaram, and J. Aarthi. (2015). Randomized Fault-Tolerant Virtual Backbone Tree to Improve the Lifetime of Wireless Sensor Networks. *Computers & Electrical Engineering*, 48, 286–297, ISSN:0045-7906, doi:10.1016/j.compeleceng.2015.02.017.

22. Sakthivel, M., J. Udaykumar, and V. Saravana Kumar. (2019). Progressive AODV: A Routing Algorithm Intended for Mobile Ad-Hoc Networks. *International Journal of Engineering and Advanced Technology (IJEAT)*, 9(2), 70–74.

23. Aslam Khan, Fartukh, Wang-Choel Song, and Khi-Jung Ahn. (2019). Performance Analysis of Location Aware Grid-based Hierarchical Routing for Mobile Ad hoc Networks. *Wireless Communications and Mobile Computing*, 1–10.

24. Nair, Haripriya, P. Manimegalai, and N. Rajalakshmi. (2019). An Energy Efficient Dynamic Probabilistic Routing Algorithm for Mobile Adhoc Network. *International Journal of Recent Technology and Engineering (IJRTE)*, 7(6S3), 1699–1707.

25. Saravanan, R. (2018). Energy Efficient QoS Routing for Mobile Ad hoc Networks. *International Journal of Communication Networks and Distributed Systems*, 20(3), 372–388.

26. Gnana Prakasi, O. S., and P. Varalakshmi. (2019). Decision Tree Based Routing Protocol (DTRP) for Reliable Path in MANET. *Wireless Personal Communications*, 109, 257–270.

11 Deep Learning-Based Modulation Detector for an MIMO System

Arumbu Vanmathi Neduncheran,
Malarvizhi Subramani, Ezra Morris,
Abraham Gnanamuthu, Siddhartha Dhar
Choudhury, and Sangeetha Manoharan

CONTENTS

11.1 INTRODUCTION

Machine learning (ML) has achieved breakthroughs in many different domains, and further research is going on for the communication domain. The rising popularity of cellular networks and other devices demand high data rates and other network infrastructures. The real-time traffic increases, and the environment turns more complex and heterogeneous. Advanced ML techniques are the solution to this problem. Here, the learning of the required model is done when the trained signal is generated by extracting features and attributes from the received signal and the label is assigned through the analysis of performance for detecting. We focus on the application of deep neural networks to improve the performance of the model. In the case of a huge volume data, deep learning (DL) is used. It is possible by the combination of strong regularization techniques like dropout, which reduces overfitting issues, and high-performance graphics-card processing power, which reduces time consumption, greatly improved methods for stochastic gradient descent (SGD) and lowered the cost, like the combination of key neural network architecture innovations such as rectified linear units and convolutional neural networks (CNNs).

11.1.1 ML FOR WIRELESS COMMUNICATION

The basics of ML concepts are briefly explained, and the implementation of ML in communication systems using supervised and unsupervised data. This helps us understand the significance of ML techniques in communication systems applications (1).

In unsupervised learning, a raw sampled time-series representation of the radio communication signals is learnt. It suggests an easier way of translating communication signals in a channel using existing unsupervised learning techniques for the neural network. The growing demand of mobile traffic is tackled by using ML techniques in 5G and beyond (2).

Timothy James et al. depicted the adaptive modulation technique used for a multiple-input multiple-output (MIMO) orthogonal frequency division modulation system. The adaptability of the model is predicted by running simulations of optimal transmission modes. It is useful in our project for the second neural network to predict the system's performance using the Bit Error Rate (BER) of the signals (3).

ML algorithms are applied for adaptive modulation and coding in wireless MIMO systems. Supervised learning– and reinforcement learning–based algorithms are introduced and validated using simulations with IEEE.802.11n parameters. This has been developed to address various channel conditions, flexibility and adaptive. This method has proved to be intelligent and more reliable (4).

The data-driven prediction and optimization-driven decision algorithms are used for antenna selection in wireless communication systems. A training data set is manipulated from the channels by designing training samples and labeling them based on key performance indicators. Data learning is handled based on k-nearest neighbor (KNN) and Support Vector Machine (SVM) classifiers. Based on multiclass classifiers, antenna selection occurs. The performance of the selection algorithm in classification and communication is evaluated based on error rate (5).

State-of-the-art Internet of Things (IoT) communication networks and the challenges are detailed. The ML aspects for dealing with the physical layer and the basics for that are provided. Supervised learning, unsupervised learning and reinforcement learning algorithms for IoT communications are detailed. How ML concepts can be applied in the physical layer and adaptive physical layer concepts are mentioned. ML for upper layers and the future open research problems in them are stated in this chapter (6).

11.1.2 DL FOR WIRELESS COMMUNICATION

This chapter discusses the supervised and unsupervised techniques to accomplish this target and the limitations of using these techniques that need improvement. Why deep learning is more preferred than machine learning is explained. DL can handle huge volumes of data as input. Also, CL can accept raw input data as well as distorted data. A neural network uses more than one hidden layer. The SGD algorithm and its implementation were learned. Furthermore, many applications are of DL in wireless and network security are studied (7).

The necessity of mobile devices and the basics of DL in wireless networks is discussed. As 5G networks are designed to reduce high mobile traffic and network resources, DL is introduced in mobile networks. DL can help manage high volumes of data and will also need fewer algorithm-driven applications. There still some limitations that should be researched for its successful implementation (8).

A conditional generative adversarial net (GAN) is introduced to employ the deep neural network completely on an end-to-end wireless communication system. The conditional GAN works on the channel conditions so it can connect the transmitter and receiver neural networks. And when the transmit signal sequence length is long, convolutional layers are used to address the issues. Hence, without complete information about the wireless channel, the deep neural network methodology is used in the complete wireless system. Different fading channel environments, such as additive white Gaussian channel, Rayleigh fading channel and frequency selective fading channels, are modeled using GAN and are trained to prove the efficiency of the method (9).

The importance of wireless communication is explained. The adaptive schemes transmit antenna selection and opportunistic relaying are discussed. The traditional channel prediction schemes based on statistical, parametric methods and autoregressive methods are detailed and introduced recurrent neural network (RNN)–based channel prediction schemes. These RNN-based channel predictors are effective under an outdated channel state information condition. This prediction can be applied in different fading environments (10).

An online channel state information prediction scheme named OCEAN is introduced. It predicts with the learned data samples of the other factors which affect the channel state information (CSI). A combination of convolutional neural network and long short-term memory is used to train the network. The prediction and accuracy of the method are validated using experimental results of different environments (11).

The complex heterogeneous network environment, unknown channel conditions and channel-related information and the requirement of high speed and high data rates are key aspects of next-generation networks. This chapter deals the mentioned challenges with DL methodologies. DL used for recognition of modulation, decoding the channel and detection. Auto encoders based on DL to deal with end-to-end communication systems are explained in detail (12).

11.1.3 ML in Various Applications

Machine learning methods and their respective learning algorithms, such as KNN, SVM, principal component analysis, Q-learning, deep reinforcement learning, and others, are studied. ML applied to the management of resources such as power, spectrum, memory, and backhaul. ML for networking, localization and resource management is explained in every section with all possible learning methods. Along with this, the conditions for applying these ML techniques are detailed. The necessary alternatives are also provided for each specific discussion (13).

The integration of unmanned aerial vehicles (UAVs) to wireless networks is discussed. The contributions of ML for the enhancement of future UAVs are discussed. The deployment of UAVs, the optimization of their trajectories and using aerial vehicles as caching devices are important discussions in the chapter. There are plenty of ML algorithms that can be implemented for UAVs. Also, the integration of UAVs with the IoT will benefit future aerial vehicle communications (14).

Index modulation (IM) becomes the spectral and energy-efficient method in wireless MIMO communications, which is also discussed by authors in their previous work (15). The detector for orthogonal-frequency-division-multiplexing (OFDM)-IM system is designed based on CNNs. Error and complexity analysis has been done for validating CNN-based detector for an IM system (16).

11.2 NETWORK MODEL 1

This chapter deals with two neural network setups for the classification of modulation schemes and the selection of transmit antenna. We consider a Nt*Nr MIMO network shown in Figure 11.1 consists of a transmitter with Nt transmit antennas and a receiver with Nr receive antennas. The information bits $(u_1, u_2, \ldots, u_b)$ are modulated using PSK/QAM modulation of modulation order M. The modulated symbol $(m_1, m_2 \ldots, m_{\log2(M)})$ is transmitted through the selected antenna over a Rayleigh fading channel over the air signal is observed at the receiver.

The received signal (y) is given as input to the neural network. The neural network predicts the signal modulation (M) and the BER performance of the system. Also based on the channel state information (CSI) observed at the receiver, the neural network matches the decision of which transmit antenna is selected (N_{ts}) for transmitting the modulated symbol (m_x).

11.2.1 CREATION OF MIMO DATA SET 1

Massive MIMO and deep MIMO data sets have been created using simulations, and hardware experiments are available in online. The radio ML group has released data sets of 24 different modulations over fading channels for a range of signal-to-noise ratio values. This network model is trained using radio ML data sets. First, we used the available radio ML data set and extracted the necessary data points for the required samples. The modulations types considered for training the network are labeled for ease of representation in the network (17–18).

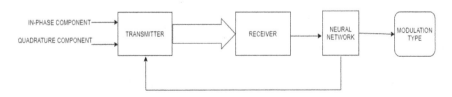

FIGURE 11.1   System Model for Neural Network 1.

	A	B	C	D	E		DW	DX	DY	DZ	EA
1		0	1	2	3		125	126	127 SNR		ModType
2	0	0.192848	-0.29562	-0.80536	0.202395	0.8(	-0.9187	0.621881	0.140654	-4	0
3	1	-0.07263	0.775924	-0.1154	-0.00189	-0.(	-0.03951	-0.56285	0.027584	-4	0
4	2	-0.32746	-1.17557	-0.1042	-0.65009	0.0(	-0.43326	-0.36013	-0.30849	-4	0
5	3	0.709679	-0.42135	-0.19336	0.232393	0.8:	-0.8509	0.352321	0.258637	-4	0
6	4	0.100531	0.037706	0.139474	0.472366	0.7:	1.172613	2.398135	1.281176	-4	0
7	5	-0.17002	-0.37069	-0.4388	-0.65128	0.4-	-0.93009	-0.13122	0.547938	-4	0
8	6	-0.37326	0.439865	0.283585	1.369873	-1.(	-0.7095	0.135434	-0.27627	-4	0
9	7	0.230244	-0.17534	-0.78509	-1.27448	0.3:	-0.7799	-0.8885	-1.0349	-4	0
10	8	0.095635	-0.31106	0.352497	0.287485	-0.`	1.008523	-1.03865	0.526623	-4	0
11	9	-0.38269	-0.18743	0.196878	0.46347	0.4(	-0.70304	-0.41667	0.115664	-4	0

FIGURE 11.2 A Portion of Data Set 1.

The modulated symbol is transmitted through a Rayleigh fading channel. The system channel impairments, such as attenuation, are considered. In-phase and quadrature-floating points belong to different types of PSK/QAM modulation. From the radio ML data set, we take 128 sample points of modulations 16PSK, 32PSK, 16QAM and 32QAM. These sample points are observed as simulation results of MIMO system of signal-to-noise ratio (SNR) ranging from −20 dB to 30 dB.

The data set can be formed as two cubes, L*S*I and L*S*Q,
where L is the number of samples,
S is the number of modulations considered and
I and Q are the in phase a quadrature points of selected samples.

The sample points are saved as. csv file and imported as input for neural network. All the simulation values are stored in Excel, and a portion is shown as an example in Figure 11.2.

11.2.2 Training the Neural Network

A multilayered perceptron is a logistic regression classifier whereby the input layer is transformed using a nonlinear transformation by learning the inputs and then mapping the data points onto a set of appropriate outputs. It has three main layers: the input layer, the hidden layer and the output layer. The training algorithm used is SGD algorithm. This algorithm calculates the derivative from each training data instance and updates the calculations immediately. This algorithm is used to transform from the input layer to a different hidden layer and then to the output layer. The output of the network is compared to the expected output, and an error is calculated; it is propagated back using back-propagation algorithm. The model is initiated by defining the layers and activation functions. The activation function introduces nonlinearity to the model so that can be mapped with real-world data. The activation function used in this network is a rectified linear activation function (ReLU). The network is forward-propagated, and the output is calculated. Gradients are calculated for the various parameters in comparison with this loss, and the parameters are then adjusted through back propagation using gradient descent. A neural network takes the input and output points, learns the function between them and predicts the outputs. Here, the inputs are the floating points, and the output is the modulation type.

The network shown in Figure 11.3 consists of one input layer consisting of 128 bits, which are the 128 samples of the floating points followed by three hidden layers, which are transformed into 64, 32 and 16 data points using the SGD algorithm. The output layer has 4 bits, and the most probable output is selected as output. The activation function is an ReLU function. The ReLU is a piece-wise linear function. If positive it will drive the input as output directly otherwise, it will drive the output to be zero. It is defined as: f (x) = x+ = max(0, x)

The epoch, that is, one cycle through the full training data set, is 10000. The step size at each iteration while moving toward a minimum of a loss function, which is said to be the learning rate, is 0.1. The code for the neural network is written in Jupyter Notebook using python3 and a front-end application is made in Flask. The network is trained using PyTorch, which is an open-source ML library.

The trained neural network output is displayed in front end an example is shown in Figure 11.4. Since the 128 samples cannot be typed as the input, so each index number is assigned for a set of 128 floating points. The output modulation number is displayed as the output of the system.

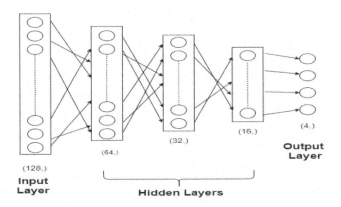

FIGURE 11.3   Architecture of the Neural Network.

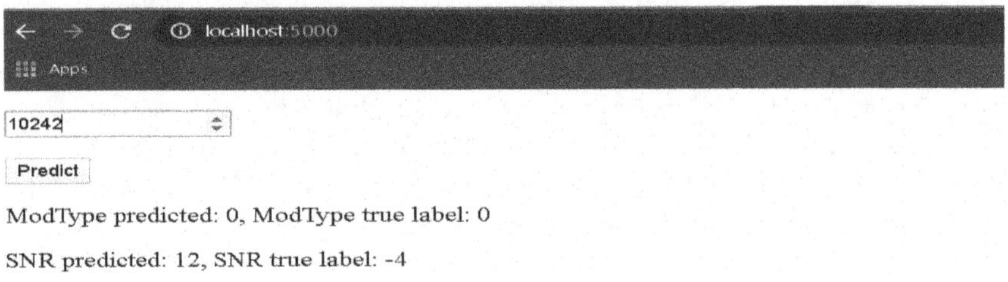

10242

Predict

ModType predicted: 0, ModType true label: 0

SNR predicted: 12, SNR true label: -4

FIGURE 11.4   Front-End Display of Network Output.

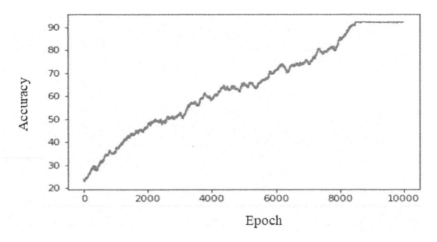

FIGURE 11.5   Epoch versus Testing Accuracy for Network 1 Modulation Type.

The accuracy, that is, the number of times the actual value is close to the predicted value of the neural network, can be improved by training with huge amounts of data and a large number of repetitions of the training algorithm. Figure 11.5 shows that the accuracy of the neural network improves with epochs.

In Figure 11.5, the accuracy value is depicted as 91.43 when a training network reaches 10,000 epochs. So it can be inferred as the modulation type predicted is 91.43 percent correct.

11.3 NETWORK MODEL 2

The network shown in Figure 11.6 consists of a MIMO system with 40-bit input, modulated with QAM/PSK modulator. Information bits $u_1, u_2, \ldots u_b$ are modulated by PSK/QAM modulation of order M. The modulated symbols $m_1, m_2, \ldots, m_{\log2(M)}$ are transmitted through a Rayleigh fading channel of channel matrix Nr*Nt. The transmitter selects one particular antenna, $N_{t(s)}$, for the transmission base of neural network predictions of performance. That is, the transmission antenna that could

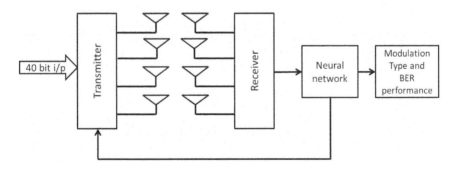

FIGURE 11.6   System Model for Neural Network 2.

result in better error performance is selected for further transmission of modulated symbols. The received signal points of samples are states in. csv file and imported as input for the neural network. The role of the neural network is to predict the modulation and choose the transmit antenna for better performance.

Apart from other detectors, this ML-based detector predicts the modulation and demodulation of the information precisely. If we train the neural network with more and number of samples, the precision and accuracy can be improved. In this chapter, we considered 40-bit digital signal information for 16QAM and 32QAM modulations for 10 different combinations. Considering a greater number of modulation techniques requires high storage devices. For simplicity and testing the neural network accuracy only two modulation techniques are taken into account.

The BER of all possible channels in the 4*4 channel matrix is calculated. This will guide in the transmitter operator selection. The received signal is y = h * x + n, where h is the Rayleigh fading parameter, x is the transmitted signal and n is the additive white Gaussian noise (AWGN) parameters.

11.3.1 CREATION OF MIMO DATA SET 2

For the second neural network, the data values are added as the result of a simulation of MATLAB code for calculating the BER of 40-bit digital signal for 16 QAM and 32 QAM for different combinations. As the consideration of more modulation techniques requires high storage devices, only two modulation techniques are considered. The BER is calculated for the MIMO system with a dynamic Rayleigh channel matrix for all 16 channel indices as shown in Figure 11.7.

The label '0' indicates channel from transmitter 1 to receiver 1, label '1' indicates channel from transmitter 1 to receiver 2. Likewise, all possible channels of this 4*4 system are labeled and used in the prediction. Here, for each combination of input bits, the dynamic channel environment is considered in the simulation, and the

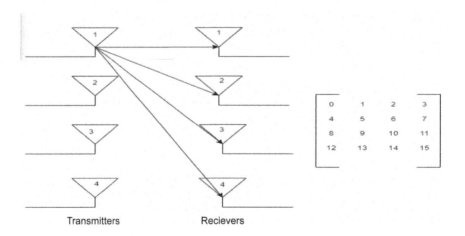

FIGURE 11.7 System Model and Channel Matrix for Data Set Creation of the Second Neural Network.

output resulting signal points are stored in an Excel sheet, which is later extracted by the neural network. The error performance of MIMO system is measured based on above received values.

The BER is predicted as the ratio of the number of error bits by the total number of bits. For analyzing the performance of system, if the BER is less than 0.05 (maximum of 2 error bits in 40 bits), the channel is efficient; else, the channel is inefficient, and the condition is checked for different channel indices.

11.3.2 TRAINING THE NEURAL NETWORK

The second neural network in Figure 11.8 is also created using a multilayered perceptron. It consists of one 43-bit input layer followed by two hidden layers followed by one output layer of 1 bit.

The input layer consists of 43 bits: 40 bits for the input digital signal, 1 bit for the modulation type, 1 bit for the channel matrix and 1 bit for the channel index number. The hidden layer consists of 20 bits and 10 bits transformed using SGD. The output layer consists of 1 bit that predicts the BER. The activation functions for the first two layers are an ReLU function, which is used when a positive value requires an output, and that of the output layer is a sigmoid function, which is used when the output is a probability function. The sigmoid function limits the output to a range between 0 and 1. The epoch is 10,000, and the learning rate is 0.1.

After the BER values are predicted by the network, a front-end application for checking the condition for efficiency of the channel on the basis of BER value is made using Flask, and the performance is displayed.

The selection of transmitter and receiver is labeled by the channel index number as shown in the matrix in Figure 11.7. With the information bits input and channel parameters, the neural network predicts whether the channel is efficient or not.

Because the input of the system is 40 bit and can't be entered in the front end, each combination of input bit is assigned an input index. Along with the input index, the modulation type, the channel number and the channel matrix are given as input. The output of the network is the performance of the channel. Because the second neural

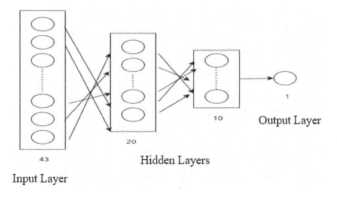

FIGURE 11.8 Architecture of Neural Network 2.

network is a regression-type example, we cannot calculate its accuracy. The performance is analyzed using the graphs depicted in Figure 11.9.

The BER predicted by the neural network, and the BER calculated by maximum likelihood detector is plotted in Figure 11.9. It shows the prediction is close to actual values.

The graph in Figure 11.10 is used to check the overfitting and underfitting. Overfitting is a modeling error that occurs when a function is too closely fit to a

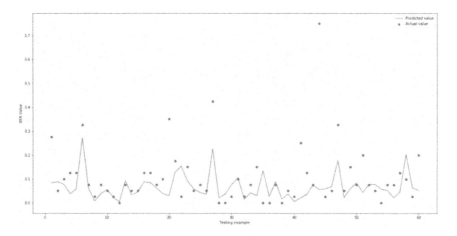

FIGURE 11.9   Predicted BER versus Actual BER of Network 2.

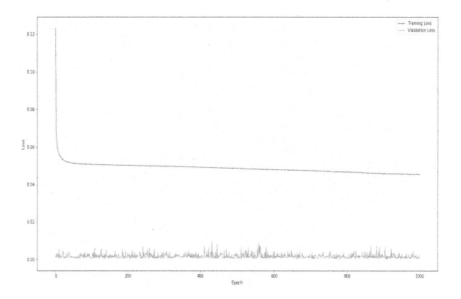

FIGURE 11.10   Loss versus Epoch for Network 2.

limited set of data points. Underfitting refers to a model that can neither model the training data nor generalize to new data.

11.4 CONCLUSION

The neural networks are created using multilayer perceptron and trained using an SGD algorithm and activation functions for the purpose of modulation classification and antenna selection. Modulation classification and channel characteristics prediction are performed by training the neural networks. These networks for classifying the modulation type and analyzing the system's performance based on BER are designed and trained.

It is an efficient way to select a better communication system that can be used to cope with the growing demand for high data rates and more automation in the future. There are many realistic constraints for implying this model in real life, but with the advancement in DL techniques, new applications and methods are being developed to implement the communication system.

DL techniques make it easier to train huge amounts of data using fewer algorithms. It is also a more comfortable way, compared to the manual techniques, for analyzing the channel characteristics. These are comparatively cost-efficient but will require skilled people to operate them. These advanced ML techniques will be used in 5G and beyond.

11.5 ACKNOWLEDGMENT

This research work is supported by financial assistance provided by a Council of Scientific and Industrial Research—Senior Research Fellowship.

REFERENCES

1. Simeone, Osvaldo. (2018). A Very Brief Introduction to Machine Learning with Applications to Communication Systems. *IEEE Transactions on Cognitive Communications and Networking*, 4(4), 648–664, 21 November.
2. Timothy, J., Corgan J. O'Shea, and T. C. Clancy. (2016). Unsupervised Representation Learning of Structured Radio Communication Signals. *2016 First International Workshop on Sensing, Processing and Learning for Intelligent Machines (SPLINE)*, 1–5, 6 July.
3. Timothy, James O'Shea, T. Tamoghna Roy, and Charles Clancy. (2018). Over-the-Air Deep Learning Based Radio Signal Classification. *IEEE Journal of Selected Topics in Signal Processing*, 12(1), February.
4. Zhang, L., and Z. Wu. (2020). Machine Learning-Based Adaptive Modulation and Coding Design. *Machine Learning for Future Wireless Communications*, 157–180, 3 February.
5. Joung, J. (2016). Machine Learning-Based Antenna Selection in Wireless Communications. *IEEE Communications Letters*, 20(11), 2241–2244, 27 July.
6. Jagannath, J., N. Polosky, A. Jagannath, F. Restuccia, and T. Melodia. (2019). Machine Learning for Wireless Communications in the Internet of Things: A Comprehensive Survey. *Ad Hoc Networks*, 93, 101913, 1 October.

7. Eugenio Morocho-Cayamcela, Manuel, Haeyoung Lee, and Wansu Lim. (2019). Machine Learning for 5G/B5G Mobile and Wireless Communications: Potential, Limitations, and Future Directions. *IEEE Access*, 7.

8. Zhang, C., P. Patras, and H. Haddadi. (2019). Deep Learning in Mobile and Wireless Networking: A Survey. *IEEE Communications Surveys & Tutorials*, 21(3), 2224–2287, 13 March.

9. Ye, H., L. Liang, G. Y. Li, and B. H. Juang. (2020). Deep Learning-Based End-to-End Wireless Communication Systems with Conditional GANs as Unknown Channels. *IEEE Transactions on Wireless Communications*, 19(5), 3133–3143, 6 February.

10. Jiang, W., H. Dieter Schotten, and J. Y. Xiang. (2020). Neural Network—Based Wireless Channel Prediction. *Machine Learning for Future Wireless Communications*, 303–325, 3 February.

11. Luo, C., J. Ji, Q. Wang, X. Chen, and P. Li. (2018). Channel State Information Prediction for 5G Wireless Communications: A Deep Learning Approach. *IEEE Transactions on Network Science and Engineering*, 7, 25 June.

12. Wang, T., C. K. Wen, H. Wang, F. Gao, T. Jiang, and S. Jin. (2017). Deep Learning for Wireless Physical Layer: Opportunities and Challenges. *China Communications*, 14(11), 92–111, 22 December.

13. Sun, Y., M. Peng, Y. Zhou, Y. Huang, and S. Mao. (2019). Application of Machine Learning in Wireless Networks: Key Techniques and Open Issues. *IEEE Communications Surveys & Tutorials*, 21(4), 3072–3108, 21 June.

14. Klaine, P. V., R. D. Souza, L. Zhang, and M. Imran. (2019). An Overview of Machine Learning Applied in Wireless UAV Networks. *Wiley 5G Ref: The Essential 5G Reference Online*, 1–5, 30 October.

15. Neduncheran, A. V., M. Subramani, and V. Ponnusamy. (2018). Design of a TAS-STBC-ESM (F) Transceiver and Performance Analysis for 20 bpcu. *IEEE Access*, 6, 17982–17995, 3 April.

16. Wang, T., F. Yang, J. Song, and Z. Han. (2020). Deep Convolutional Neural Network-Based Detector for Index Modulation. *IEEE Wireless Communications Letters*, 9(10), 1705–1709, 11 July.

17. Mao, Qian, Fei Hu, and Qi Hao. (2018). Deep Learning for Intelligent Wireless Networks: A Comprehensive Survey. *IEEE Communications Surveys Tutorials*, 20(4), fourth quarter.

18. Ha, Chang-Bin, Young-Hwan You, and Hyoung-Kyu Song. (2018). Machine Learning Model for Adaptive Modulation of Multi-Stream in MIMO-OFDM System. *IEEE Access*, 7, 21 December.

12 Deep Learning with an LSTM-Based Defence Mechanism for DDoS Attacks in WSNs

P. Manju Bala, S. Usharani, T. Ananth Kumar, R. Rajmohan, and P. Praveen Kumar

CONTENTS

12.1 INTRODUCTION

Wireless sensor networks (WSNs) have recently been implemented in a wide variety of uses and play a key factor in the current area of science (1). Information has been developed; it has currently become a major component of the Internet of Things (IoT)

(4). With the wireless radio transmitter, the WSNs temporally involve distributed small, low-power sensor devices to detect the different physical processes and analyse the information in all kinds of situations. If the certain and wireless Local Access Networks (LANs) are contrasted, the WSNs may respond to an intense scenario. Data processing and collaboration between one of the base station sink nodes are involved (2). Thus, the wild biome protection (5), real-time manufacturing control and robotics (3), construction security observation, traffic monitoring and management (5), ongoing health monitoring (6), and army applications (7), among others, are commonly used in several civilian applications. Because of their restricted power, arbitrary distribution, and unwanted activities, the sensor nodes are vulnerable to another assault, and their protection is jeopardised in a serious situation such as adverse zones (8). WSNs are especially vulnerable to denial-of-service (DoS) attacks during deployment in a hostile environment, where the sensor nodes are actually seized and manipulated (9). Defence methods, including authentication, used in wireless devices, owing to the similarity of sensor nodes, could not be used on WSNs. By communicating with data streams, an attacker can restrict a WSN through the transmission of black hole attacks, selective forwarding attack, identity replication attack, wormhole attack, Sybil attack, and HELLO flood attack (10).

The network capability is reduced or removed, and Distributed Denial of Services (DDoS) assaults (11) decreases or reduces terminations of ordinary communications by flooding. DDoS attacks in WSNs typically vary from other wireless LAN networks. Nearly all layers in WSNs are subjected to a number of DDoS attacks and have varying attack methods (12). Many ideas are built to keep the network safe from a DDoS attack. The DDoS attacks on various open systems interconnection layers in WSNs and security tactics are defined. Watchdog systems typically classify the misbehaviour of nodes and use the path without malicious nodes for rerouting. According to a single node neighbour's position, selfish nodes can be graded in the Credibility Rating Scheme (13), energy use and packet delivery, for instance. WSNs use machine learning algorithms or a signature-based algorithm from Beacon. One of the main (14) criteria used in these position algorithms is the Obtained Signal Strength Indicator, which never implements node algorithms. The virtual currency mechanism stops consumers from overwhelming the information-sharing mechanism with a credit/virtual fee (15). Wood and Stankovic suggest a protocol for identifying and visualizing the jammed area in a sensor network using a mapping protocol (16).

The latter mechanism does not offer a detailed solution. The moving area substation solution proposed a defensive device to identify and remove WSN DoS attacks (17). The least mean square distinguishes malicious communications, which essentially reduces energy use and system consumption and improves message self-authentication. Finally, on its intra-cluster architecture, the cluster head (CH) authenticates and organizes its member nodes to alleviate DoS risks and improve network stability (24). Masked Authenticated Messaging (MAM) protocol is a multi-agent protocol of communication that is compliant with the multi-user method on a single device. Thus, the data-sharing mechanism is partial to the number of users running the system at the given period. The output is restricted, given that the system agents' requests to exchange goals or facilitate sharing functions through some other variables (22). This strategy would be used by CH to distinguish contaminated data from real messages sent to the base station (BS). Cluster organization and MAM recognition are also given and deliberated. The discourse of effects is presented. This technique does not, however, hit the WSN capital quota (21). Moreover, node-level threats are not specified. This chapter suggests a WSN system to identify and isolate the extensive DDoS attacks on the basis of a deep learning (DL) algorithm. A wide-ranging anomaly detection assessment is provided in WSNs (20); DDoS attacks are, in comparison, mainly targeted at depleting the capacity, resources, and defences of the deployed nodes. The jamming detection mentioned by Adimoolam et al. (18) for detecting attacks at different levels of DDoS is used and analysed. The system provides an incentive, through a worldwide system and controller object, to strengthen our network security (19). A DL technique is used to identify flow-related anomalies based on a recurrent neural network (RNN) model to minimise possible threats. This chapter discusses long short-term memory (LSTM)-based profound learning to achieve energy usage and optimised load balance in the WSN fusion core. This chapter also includes spatial knowledge focused on the DL that helps localise and identify different threats. In this algorithm, the criteria are strictly enforced; the limits generated by intruders' intrusion must be identified within a given time. The attackers are then identified, and counter measurements are occupied using a data detector query handling (23). In the proposed system, almost the same efficiency is obtained if the alternative positions are being used instead of conventional localizations. In WSNs, the overhead of the fusion centre is significantly reduced with the LSTM model. Simultaneously, the energy necessary for LSTM data broadcast is significantly smaller than the energy used for real data transmission. The DL LSTM model has been used to experiment with a wide range of occult layer nodes and indicating cycles. The investigational result showed that the LSTM decreases energy consumption, overhead signalling, average suspension, and maximises total performance relative to other approaches.

12.2 PROPOSED MODEL

The proposed DDoS detection architecture is investigated and described in Figure 12.1a. In a WSN, relatively homogeneous small sensor nodes with identical energy levels work together to gather data from their IoT fields. Each sensor node has wireless interfaces to relay collected data. The sensor node modifies its values, such as responsibility period and CH parameters, as complex as sensors. All sensor nodes are developed and uniformly located within a region of 1100 m × 1100 m, while every sensor node covers a range of 150 m. All nodes in the network are known as stable.

The WSNs are distributed, depending on the network's properties, into groups or clusters. The cluster heads are chosen by choosing high outstanding energy nodes between the neighbours. If the CH's outstanding energy falls below the thresholds, the CH leadership election method will be triggered. The selected CH acts as the node of its cluster or intermediate nodes. Considering the propagation of various network access with N bands from the CH to the dynamic data separation cluster

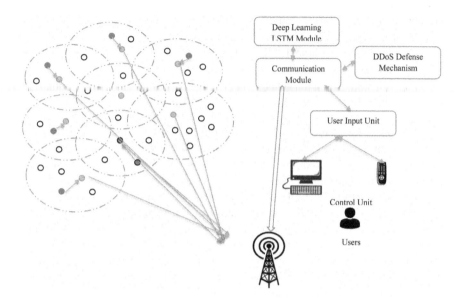

FIGURE 12.1a DL LSTM-Based Defence Mechanism Design for Attacker Detection.

nodes (subcarriers), the N bands, with aid of the CHs, are distributed among their group nodes (Figure 12.1b). Spectrum sensing is done regularly by CHs using before the subcarriers to relay the data to their community nodes.

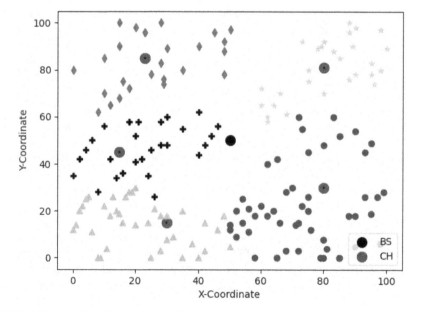

FIGURE 12.1b Clustering of Sensor Nodes.

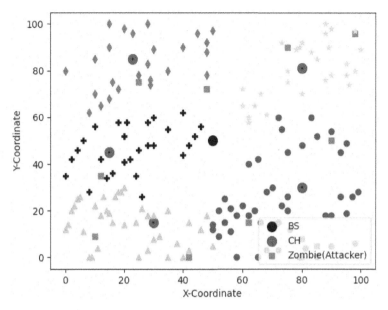

FIGURE 12.1c   Clustering of Sensor Nodes.

The attack detection module will search the design vectors, if the CH is now involved in any other contact and in-between if entry to the particular CH containing it is allowed, or if no records are authenticated, then the requisition is made to remain in the block list (Figure 12.1c). Our DL LSTM-Based Defence Mechanism (DLLDM) algorithm identifies the node activity as natural or suspicious on the basis of the criteria collected signal power, packets received per minute, receiving packet suspension, information shift, information falling ratio, and information forwarding ratio. It will be further included to the node in the block list if listed as harmful. So the target node will be tested in the mode of the block list further during the information-forwarding process (IFP). When the node in the block list is identified as visible, the packets will be dropped. In a cluster, based on its transmission distance, the CH requires power control to preserve active signal power levels for each node. Designing a stable network will avoid the energy usage of calculating resources such as bandwidth, battery and the process time. The primary reason is energy utilization of the node during processing time. The following are some significant advantages of CH selection in WSN:

- When a WSN that involves a BS and a sink with the presence of self-organizing and self-configuring nodes is of concern.
- The lifetime of the sensor is typically extended by the adequate choice of the routing protocols.
- The outcomes of the energy usage of the sensor node are focused on the CH and the BS during intervention studies.
- By avoiding various kinds of threats on the BS and the CH, lifetime enhancement of the network can be accomplished.

12.3 LSTM PRINCIPLE

The principle used in this chapter is the LSTM model. Infer that there is a fusion core in a WSN separated into two phases (Figure 12.2a) and three layers of LSTM (Figure 12.2b). It is mentioned that both the WSN topology and the LSTM framework are hierarchically the same. A potential approach is available that enables the splitting the layers of LSTM and the assignment of different WSN stages. A WSN is known to have been divided by n stages, and m layers are present in LSTM. Each part is fulfilled by measuring the computing units in the related stage of hu in the WSN, thus splitting m layers into k measures (k ≤ n, m), and the LSTM will be illustrated in the coming sections.

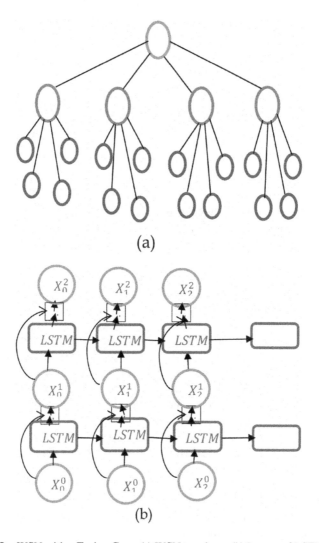

FIGURE 12.2 WSN with a Fusion Core: (a) WSN topology, (b) Layers of LSTM.

12.3.1 LSTM Model

The LSTM model is used to overcome the problem of decreasing RNN gradients in DL. It acts like a regular RNN along with a hidden layer, in which a memory cell can be used to replace any usually hidden layer node as shown in Figure 12.3. By ensuring that the gradient is able to move over many times without removal, the memory cells consist of a recurrent edge that is self-connected. Subscript c has been added to evaluate certain memory cell references and not an ordinary node. In terms of weights, a simple RNN has long-term memory. At the time of training, it is progressively changed, encoding general data knowledge. To establish an ephemeral function, it is also consisting of short memory, which transfers from any single node to consecutive nodes. The middle form of the memory cell is denoted here by the LSTM technique. A memory cell is known to be a composite group, generated in a specific connectivity process from ordinary nodes by including the multiplicative nodes indicated by letter $\prod$ in the figures. Each LSTM cell unit is enumerated and defined as stated in the following discussion. For example, at each memory cell c of a layer, s represents a vector with a value of i_c. If subscript c is used, then indexing a single memory cell helps.

Input: This is labelled as u_c. It is a node which attains the start from a usual method of input layer l(x) in a current time interval and from a hidden layer at the previous interval step, h(x − 1). Frequently, the added input weight is a tanh function activation, still it results from the LSTM, and function activation is represented by a sigmoid (σ).

Input unit: These units are anticipated to be distinctive functions of the LSTM model. These units look like a sigmoid of input, which takes the stimulation from recent data point l(x) and the hidden layer at an earlier time period. It is named as a unit due to the increased value of another node. When the value is set to zero, it is referred to as a unit in which the flow of various nodes is cut off. Each flow is recognized when it is recognised as one. In addition, the value of the u_c input unit increases the value of the node's input.

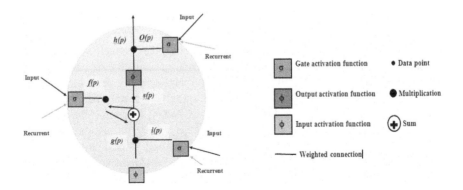

FIGURE 12.3   LSTM Memory Cells

Inner state: The centre point of all memory cells is node i_c with constant stimulation, called in the specific papers as the several internal states of the cell. The internal state i_c was involved with the self-connected regular edge along with a constant unit weight. In the existence of an explosion, error flows over the clock cycles because of the edge of neighbouring clock cycles of comparable weight. It is designated as a carousel of persistent error. For vector function, the modernisation for the internal state is

$$s^\wedge((x)) = g^\wedge((x)) \odot u^\wedge((x)) + s^\wedge((x-1)) \tag{12.1}$$

where $\odot$ specifies multiplication process.

Forget unit: The unit g_c has been presented. It gives an atypical model the ability to study the framework of the inner state. It is further appropriate in repeatedly running networks. By forget units, the function for evaluating the forward pass in internal state is

$$s^\wedge((x)) = g^\wedge((x)) \odot u^\wedge((x)) + f^\wedge((x)) \odot s^\wedge((x-1)) \tag{12.2}$$

Output unit: The i_c value is created by the memory cell, which is known to be the inner-level i_c measurement enhanced by the resulting gate r_c value. In particular, the internal state is typically enforced by the tanh activation function, because it includes simulated results of each cell with a dynamic range comparable to that of the rapid tanh hidden unit. In an alternative neural network study, however, determined linear units are composed of higher dynamic ranges that are easy to train. It is also possible for nonlinear functions.

Although the LSTM was initiated, several variations were found. In 2000, forget gates, described earlier, were presented but not part of the actual suggested LSTM. But in most modern executions, they have been verified as successful and are normal. The peephole relationships that pass immediately from its internal level to the input and output units of the associated node that neglects primary ownership to be modified to the output gate. They clarify that these links improve action on time positions in which the node can perform calculating specific breaks between stages. With the corresponding example, the insight of the linked peephole is grasped. Infer a network in which an object, if seen, can acquire to measure objects and produce few necessary results. The network will discover to allow different suitable complete activation in the internal state behind every item seen. These inputs are recorded with a constant error carousel in the internal state i_c that is iteratively improved every period until another entity is realized. If the nth object is realized, the system needs to be defined to release data from the internal state such that the result is included. To do this, the result gates r_c have to know that the internal state i_c is fulfilled. So i_c has to be an input for r_c. In the LSTM model, the estimation is based on memory cells and is configured properly. At any point of the time, the corresponding calculations were carried out.

At time x, the hidden layer value of the vector in the LSTM is h(x), as h(x −1) is the result of each memory cell at the corresponding time in the hidden layer. Remember that these calculations include, although not peephole connections, the forgotten gate. With setting f (x) = 1 to any x, computations are achieved to make the LSTM easier without forgetting gates. Following the modern style, we use the tanh value φ,g for the input node. The activation function is sigmoid σ for g. Unexpectedly, if allowed to activate in the internal state, the LSTM is learned based on the forward pass. Similarly, the result gate determines when to let out the value. The activation is recorded in the memory cell if both gates are closed, neither upward nor decreasing nor disrupting the result at in-between time stages. The constant error container causes the incline to spread through many time stages with regards to the backward pass, nor does the vanishing burst. In this sense, regarding whether and when to let faults in and let them out, the gates are learning. In practice, as connected to simple RNNs, the LSTM has shown a greater capacity to learn long-range dependencies.

The method in which several memory cells are used with each other to form the secret layer of a functioning neural network is a recurrent source of misunderstanding. The result from each memory cell passes with every gate of all memory cells to the input node in the following time step. It is common to have several layers of memory cells. Typically, at the same time level, each layer gets feedback from the layer below in these models and from the related level in the previous time stage.

12.4 LSTM TRAINING MODEL

12.4.1 Training the DL-Based LSTM

It is crucial to train the LSTM at the initial level in the fusion centre by conducting LSTM. Training data are an example of all WSN nodes, and the qualified LSTM values are directed to the assigned LSTM stages in different calculating units. Although a WSN is supplied with a form of training data set, this knowledge often exhausts the capacity of several networks. In reality, the data of a sensor's orders do not change in a short period. So pick one of the instances for LSTM preparation. The problem is that if knowledge shifts, we do not know it. The random approach chosen by data has been checked to be useful in resolving this problem, and a digital ID study found that better results are obtained by an arbitrary choice of 10 percent preparation. Thus, the random approach chosen is to effectively minimise the collection of duplicate data to be transmitted. Then, through the random selected technique as pursuits, we provide the training process:

1. The sensor node randomly produces a sensor identification and passes a message to the sensor.
2. The selected senor finds the message and spreads the request data.
3. Sensor node acquires the trained data from the selected senor and transfers the information to the G_{lt} method.
4. The G_{lt} method confirms if the training result acquires the end state. When OK, move to Step 5. Else, go to Step 1.
5. The sensor node transfers all portions of the LSTM organization information to the corresponding calculating unit.

The LSTM delivery hierarchy is based on its features. But with power utilisation, certain transmission systems need to be managed. The theory of planning the system of transmission relies on the trade-off between the use of computing and broadcast power. Assume that there is a computer machine and that to terminate the function, it performs c training. All training uses power from F_u. Also, without any impact or reduction, the computing unit uses E_t power to transfer a bit to the cluster head. Troubling and modulation effects lead to additional power utilisation of E_0 with every impact. First, when the following formula is fulfilled, it is stated that a processing unit is permitted to use the LSTM part:

$$cE_i \le (b_i - b_0)(E_t + E_0), \qquad (12.3)$$

where b_i mentions the size of the computing unit's input in b_{it} and b_0 is the scope of the computing part's outcome in b_{it}, $b_i \ge b$. When E_0 is set to 0, next it contains

$$\frac{c}{b_i - b_0} \le \frac{E_t}{E_i}. \qquad (12.4)$$

Remarkably, when Equation 12.4 is reached, Equation 12.3 should also be met. The upper limit computational function obtained by the processing unit is resolved.

12.5 FRAME STRUCTURE

In each layer of a WSN, there will be various types of DDoS attacks, even though the network infrastructure is very carefully constructed. If the level of their intrusion (18) is lower, the attackers may not be detected. It also generates few inconsistencies during attackers' intrusion and energy inefficiency due to contact disruption (16, 36), suggesting the presence of deception in the WSN. DDoS attacks relate to the illegal act of accessing the network device because of the series of several activities of the unauthorized node. The algorithm is used to separate the attacks in this DL in the decimal floating point (27, 35). The DL with LSTM-based algorithm is used to detect attacks, even when many attackers enter the area. There are three phases: the main control system, the identification of threats, and prevention methods.

12.5.1 MAIN CONTROL SYSTEM

There are three stages in the main control system—the primary launch phase, the introduction stage of the route, and the information-forwarding stage—that can be interpreted in terms of the predictive method. Based on the sensor node's key duration and step time, public indices are loaded even prior to sensor initialization, with the possibility distribution function that can be added to decide a new separate key. The distance is established because of the vector distance and the value that is defined as a cluster between the node distances. The total life of the sensor nodes is divided into short-term durations, which are called step durations. The size of 80 bits of a public key is used based on the elliptic curve cryptography. A key length is represented as the lifetime of the key. The maximum number of keys produced depends on the lifetime of the process and the lifetime of the key. If we select a lower key period value, it will maximize the keys needed for the process, thus raising the workload. Based on a simulation model, the optimum value of the key interval is determined by changing the key interval from 4 s to 30 s. The key launch process is the basic work of the key management scheme after deployment. It is primarily used between 3 s to 8 s for connection establishing between neighbour nodes in a group and key interchange. If a key is allocated to the root node, the respective child nodes that demonstrate and satisfy autocorrelation (link creation) to the root node will be "coupled", and the link is thus generated. Two sub-stages, specifically the Bidirectional Verification Process and Key Transmission Phase, distinguish the key setup (KTP). The two-way handshake of the bidirectional verification process and bidirectional ties is known after the key sharing. Using an advanced encryption scheme, the sensor data are encoded and sent via the CH to the BS. Keys are transmitted via a community key process (17) in the key transfer phase, which is to be used to protect data in future stages. Every node maintains data around the encrypted single and two-hop neighbours in the Route Launch Process. The hashing functions can be used to authenticate the node neighbour identification sensed during the key start process. It is broken down into Single-Hop Dissemination Phase, Node Association Phase, Two-Hop Investigation Phase, and Segment Depth Transmission. The Single-Hop Dissemination Phase is responsible for the identification of single-hop neighbours and their duties. In the Two-Hop Investigation Phase, high control transmitters are similarly used to classify the two-hop neighbours. The one-way-based hash authenticated timestamp is enhanced with every node that provides synchronisation among the base station and nodes that are kept by the Node Association Phase. The IFP consists of two kinds of models for communication. All the cluster nodes sent the information in the first intra-cluster process and forwarded it to each CH. In the inter-cluster point, by the aid of other CHs, depending on the complexity level, all CHs update the data to the BS. Depending on the route setup step table, every node tracks the neighbour's behaviour, such as message data transmission, and continuously changes its depth level at periodic intervals. The IFP's primary goal is to stop DDoS attacks on the network in terms of jamming, blackhole, misbehaviour, flooding, de-synchronization, and homing attack.

12.5.2 Control Phase

This is a framework to improve user productivity, like smartphone, PC, tablet, and sensor nodes. Only via this framework can users send the requests, information packets, and further details with device log management.

12.5.3 Communication Unit

It works as the connection among the WSNs, attack detection phase, and the defence measures to avoid with the DL model that contains broadcast and response of packet data and control data constraints for recognising the attackers to dissimilar detection subsystems.

12.5.4 IDENTIFICATION OF THREATS

This is an essential phase for detecting the different kinds of attacks in the developed model by exhausting the obtainable sub-modules presented in Figure 12.4. Both sub-modules are distinct from one another and just transform their information through communication unit. The detection device interfaces consecutively obtain the requirements that describe the attack node via the communication unit in a real-world situation. Where the method is wireless and continuous, the detection framework sets the flags for each sub-module, and the related data are inferred in the contact framework. It contains many sub-modules to include defence systems, such as a process change mode, rate exertion, verification, and authentication. The defence system unit begins to avoid DDoS attacks on the network if there is an alarm from the communication unit. Based on the training process, the DL model is computer software used to track the performance and the behaviours of the WSNs. The LSTM-based DL can be used for classification that includes the testing and training process in Figure 12.5. It is a hidden layer insert scheme that incorporates the unsupervised learning model with the linear perceptron in a supervised process of executing the mapping feature. Remember that there are M inputs of length Y direction, so this neural network (38) can be denoted as

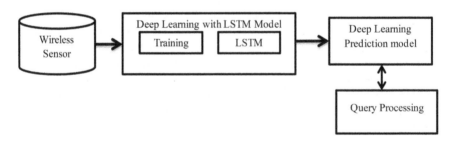

FIGURE 12.4   DL with LSTM Model.

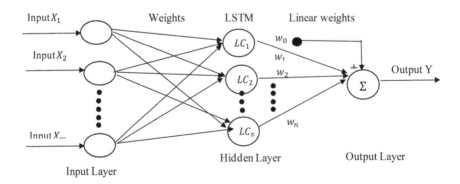

FIGURE 12.5   LSTM-Based DL Model.

$$G(Y) = \sum_{i=1}^{m} W_i \phi(Y, h_i), \qquad (12.5)$$

where h_i is the hidden neurons, h_i is the ith neuron centre, W_i is the hidden neuron weight, and LSTM feature is defined by $\pi(Y, h_i)$. The Euclidean distance between sample Y and neuron h_i centre is interpreted by a Gaussian function:

$$\phi(Y, h_i) = \exp\left(\frac{-\|Y - h_i\|^2}{2\sigma^2}\right). \qquad (12.6)$$

Typically, the training is to create the h_i neuron centre by arbitrary selection or grouping and to determine the W_i and σ variables by the process of back-propagation error:

$$\lim_{\|y\| \to \infty} \phi(\|Y - h_i\|) = 0. \qquad (12.7)$$

Using of encoding L pattern, the n source function implements a mapping with G is

$$G_i(Y) = \sum_{j=1}^{n} W_{ij}\varphi(\|Y - h_j\|) + W_{0i}, i = 0,1,2,3,4,\ldots\ldots L, \qquad (12.8)$$

where W_{0i} is the weight immersed by external neuron function $h_0 = 1$. To shorten the representation, the LSTM becomes

$$G_i(Y) = \sum_{j=0}^{n} W_{ij}h_j(Y), i = 1, i+1, i+2\ldots\ldots L. \qquad (12.9)$$

The simulation stage improves the contact and alerts the network node about the understanding of the threats before it gets infected, enabling the network to take protective methods before getting infected (test value) as if threats on other non-interest path is stopped. We have therefore followed this model of approach in this text. The "vector model" and "weight coefficients" are the key parameters of LSTM. h_j and the scaling criteria prepare and determine the system weights by the information given.

Suppose the test sets are $a\mu$ and the defined vector $b\mu$ $(a\mu, b\mu)$, $\mu=1, 2, \ldots, L$. The base function outcome with the attribute vector is denoted as $A\mu j = hj(a\mu)$. The W error is represented by the two matrices provided. $M = (M_{\mu j})$ and $N = N_{\mu j}$ is

$$E(W) = \|MW - N\|^2. \qquad (12.10)$$

The preceding calculation consuming the answer in terms of $W = M + N$, where M^+ is the matrix inverse of M, which can be defined by

$$M^+ = \underset{a \to 0}{Lim}\left(M^T M + aI\right)^{-1} M^T. \tag{12.11}$$

Depending on the conditions, they usually require updates (optimisation). The original knowledge of the message to be transmitted corresponds to this vector model, and the scalar values needed to intensify the LSTM reactions are the weight equations. For the weights, the delta learning rule is

$$\Delta W_{ij} = \eta \sum_{\mu=1}^{k} a_j\left(a^\mu\right)\left(b_i^\mu - f_i\left(a^\mu\right)\right). \tag{12.12}$$

The output in this approach is between 0 and 1 and includes Gaussian distribution depending on the consequence of equivalence among the training and testing data. Attack standards suggest that the incoherence measured in output data is measured using the values of criteria that have been met, such as the number of packets obtained per second, transmission interruptions, corruption of information, transportation and falling ratio, and so on.

Figure 12.5 will provide insight into the degree to which attacks have been carried out, the degree to which the original material has been compromised, and what type of prevention action against them should be taken (28). Input constraints used to identify the behaviour of the attacks include established signal strength, packets received per second count, packet delivery delay, are whether data changed, the data sinking ratio, and the data onward ratio—the ratio of packets forwarded and the number of packets received. In normal mode, the proportion has risen and is smaller for the intruder node.

Depending on the network output, the module selects the LSTM method of DL and machine learning. In addition, the records can be transmitted to the model, which, depending on its timetable, performs computationally costly and time-consuming processes. During the main transfer process, each and every node collects the input variables and transmits them to the CH. In the data table, the CH stores the input function data and records from the node. Data from the model are created from the training phase by adding different attack and normal network scenarios. There are three different ways to create the data sets: from a likely source, that is, the data collected from each node are saved and stored in the newly generated database; from the previous database, where it is possible to add new data; and, finally, from the child data set, where a parent data set is generated. The evaluation percentage of each data set produces the comparative data set that is imported into the database for the respective data set. The percentage calculation formula can be defined as ((recorded/ location data set) * 100). The allocation of the data set for training and testing phases is split into 70:30 combinations. Table 12.1 describes the performance indication classification of the various attackers in the results. In the training step, with the predetermined inputs and outputs, the LSTM model adjusts the centre and length of the biased function and output weights. The LSTM-based DL model is highly effective for validation; it validates 30% of data in the testing process. It is possible that

TABLE 12.1
Indication Classification

Digits	Symbol Value	Indicator Classification
000	Indicator_1	No attacks
001	Indicator _2	Jamming
010	Indicator _3	Exhaustion
011	Indicator _4	Block hole
100	Indicator _5	Misbehaviour
101	Indicator _6	Homing and eavesdropping
110	Indicator _7	Flooding
111	Indicator _8	Desynchronization

the sensor nodes will determine the vector model in response to the attacks, and the variations in the actual output will differ. Machine efficiency is based on the training, as all know the more the machine is trained will result in less vulnerability to new attacks, and similarly, the recognition of different attacks mainly rests on the training duration.

LSTM model parameters are often modified depending on the precision of the testing. Mapping, migration, and routing layers are part of the triple-layer platform.

The plotting layer is essential for choosing the appropriate open core as the first receiver in order to decrease the likelihood of jamming (40). If the exemplary is developed, the attackers in the runtime model are ready for categorization or detection. The CH now receives the record table input, transfers it to a DL model, and is categorised as regular or offensive behaviour. The CH saves the node identity in the block list if the node is marked as an intruder. The Bidirectional Authentication System (BAS) is forcefully initiated again by nodes in the block list. If encryption is fruitful, the node will stay in the system; otherwise, by broadcasting the information to all the sensor nodes, it will be removed from the future contact. The attack detection module will search the vector model if the sensor nodes or the CH is active in the present in any one of the other contact or in-between uncertainty admittance to the particular CH or sensors. Once the requester is placed on the block list, then all the features of the requester are weighed with standard legitimate elements of the respective network; otherwise, links are blocked if the connection is created. Each node can be subject to various DDoS attacks. The information obtained from sensor nodes is also considered on the basis of a DL analysis query. For example, let us assume that the receiver is receiving the data packets at different times from different sensor nodes. Once the packets are received, the DL query processor will create the query. In the existing and conventional model, once all the sensor node packets are received, the query processing will be performed, whereas a pending event will be shown before all the data from the node arrive. Here, to solve this issue, query refinement will be performed.

To assist the network with suspicious nodes, the defence system device generates several control signals. The process of communication is responsible for the transfer of response and device signals. The indicator is an alert device which, in every duty cycle, expresses the indicator vector of different protection modules. This is also a memory device that records from "000" to "111," the starting value. Similar counter values mean a different indicator. The defence counter amount unit determines the option of the several defence counter methods, based on the indicator vector value. From the eight indicators, one indicator will be set as the attack defence mechanism; it will be selected by the defence countermeasure. If the data collected from the sensor are degraded, then the ECC will be used to solve such types of problems as those listed in Table 12.1. Otherwise, in the worst-case scenario, the transmission will be resumed.

12.6 ANALYSIS OF PERFORMANCE

In the network simulator kit, the proposed system structure is simulated. Network Simulator 2 (NS2) is a separate C++-coded event simulator targeting networking science. NS2 allows full provision for Transmission Control Protocol and routing simulation. The simulation factors list is shown in Table 12.2. Approximately 200 sensor nodes are generated and positioned in an area of 1100 m × 1100 m, and the coverage area for sensor node is 110 m.

The amount of sensor nodes available in the specified network depends on the space, communications, and computing overheads. The DLLDM protocol, however, specifies that there is a variable memory allocation for any calculations can be made; that is, the storage is designated for the parameters during the processing; once the outcome is produced, the storage is eliminated and recovered for some other elements, thus resolving the problems of memory calculation.

During the indices-generating process, the amount of index produced based on the segment period and the index period (31), another challenge due to overhead. When the user chooses a small key period value, then it is reflected in an increasing number

TABLE 12.2
Simulation Factors

Sl. No	Factors	Value
1	Simulation area	1100 * 110 m
2	Node coverage range	110 m
3	Node ID	16 bit
4	Initial node energy	2 J
5	Average packet size	60 Bytes
6	Number of nodes	200
7	Simulation time	500 s
8	Transmit data rate	300 kbps

of keys to be needed for the process. It results in raising the overhead. To maintain the overhead cost so that it does not to become high, the optimal value of the main length should be between 5 s and 30 s. In the simulation scenario, eavesdropping, fatigue, and packet dropping attack models are considered. The number of nodes that strike ranges from 10 percent to 25 percent of the regular nodes. The cumulative running time for each and every model last for 189 s. The constant bit rate application is used to randomly select the source and destination pair. The data packet transmission rate typically ranges from 0.001 s to 0.01 s, and the size is about 48 bytes. The suggested model of the DDoS attack protection algorithm is organized from the parameter values.

12.6.1 Detection Performance Analysis

Different metrics can be used to find the output of a DL classifier. The efficiency of the proposed system is evaluated by measuring the following metrics.

12.6.1.1 Detection Ratio

The DDoS attack detection ratio is calculated by using the confusion matrix. It is measured by

$$Detection\, ratio = \frac{T_P}{T_P + F_N} \times 100\%. \tag{12.13}$$

12.6.1.2 False Alarm Ratio

A false alarm ratio is the amount of the proportion of normal examples unclassified as an attack. It is measured by

$$False\, Alarm\, Ratio = \frac{F_P}{T_N + F_P} \times 100\%. \tag{12.14}$$

12.6.1.3 Correctness

The correctness of the suggested method is the quantity of adequately indistinct data in the accessible data set. It is measured by

$$Correctness = \frac{T_P + T_N}{T_P + F_N + F_P + T_N} \times 100\%. \tag{12.15}$$

Figures 12.6 and 12.7 represent the efficiency of identification using a single CH. During different detection times in the 802.15.4 network, for each CH, it displays the receiver operating characteristic (ROC) curves. Here, if the total number of defenders is in the range of 10 and 20 percent, the detection ratio is 85 percent to 98 percent, and the average false alarm ratio is 15 percent. A single CH is considered. In addition, achieving a 99.99 percent detection rate and a 0.1 percent false positive ratio requires a finite period. For the actual data forwarding point, the DLLDM demonstrates a

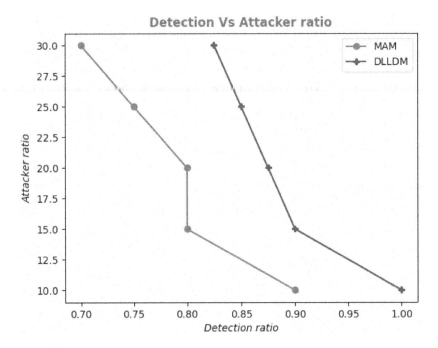

FIGURE 12.6   Detection versus Attackers Ratio.

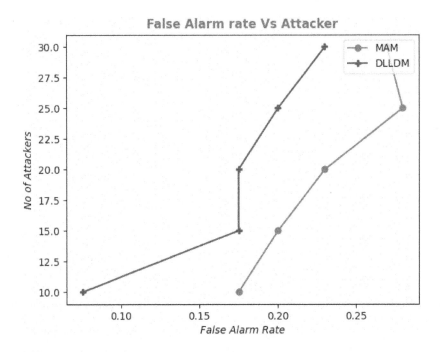

FIGURE 12.7   False Alarm Ratio versus Attackers.

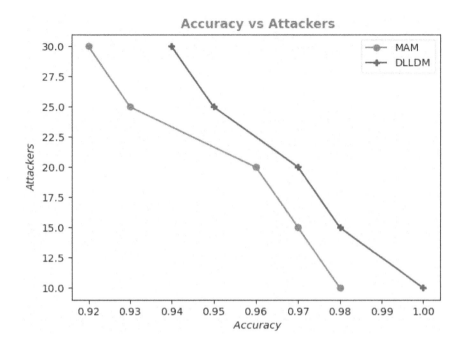

FIGURE 12.8   Accuracy versus Attackers.

difficult recognition performance as degraded false rate when compared to MAM. This gives DLLDM extra credit for the whole information sending process; the DLLDM displays higher packet delivery ratio than the MAM as the coordination of numerous malevolent systems selected as a turn pike node never harms the DLLDM. The DLLDM gives the greatest results since, if the data transmission is dropped or lost, every agent in the sending path will simply resend the information. Specifically, the DLLDM procedure functions enhance than the MAM protocol, which can provide the maximum number of network frames. The output of CHs vary similarly, and their neighbour CHs will change the worst output CHs.

Likewise, Figure 12.8 indicates the accuracy of detection is above 90 percent when the finding time is greater than 138 s, and the false-positive ratio is below 6 percent. Based on proper testing and inspection, DLLDM has the highest performance and detection level and the lowest false alarm ratio.

12.6.2 Network Performance Analysis

The performance assessment of the network is carried out in our modelling via the following parameters.

12.6.2.1 Energy Consumption

The typical energy consumption is proportional to the lifetime of the network among the min and max energy points gone over in each node. To show the diffusion of

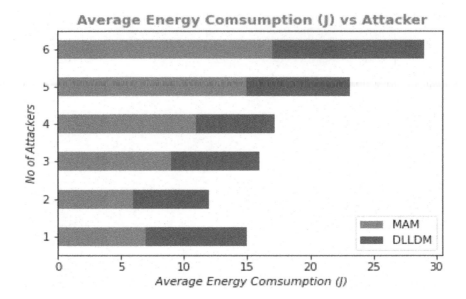

FIGURE 12.9   Energy Consumption versus Attackers.

sensor nodes by energy, a systematic analysis is performed. Based on the amount of transmitted and eavesdropped packets in the network, the attacker node packet fall levels vary, and the consumption of energy is measured. In our simulation, the network has been attacked at 200 s and acknowledged by the nodes greater than 12,000 energy points. To protect from the DDoS, our machine then continued. In Figure 12.9, because of the lower number of communication authentication phases in the forwarding direction, the DLLDM shows less energy consumption than MAM.

12.6.2.2 Throughput
Some many packets were successfully transmitted every second via the transmission medium. Figure 12.10 displays the throughput simulations against the number of MAM attackers and the suggested DLLDM solution. If the number of attackers reaches 25,000, more than 180 percent of fake packets are sent to the CHs. This improves the energy demand of CHs, resulting in a decrease in the throughput of transmissions.

The suggested DLLDM approach can accurately differentiate and safeguard against all malicious nodes and effectively remove fake messages from the network. Thus, when the DLLDM is not introduced, as the number of attackers improves, the throughput decreases.

12.6.2.3 Packet Delivery Ratio
The packet delivery ratio is described as the proportion among the target data packet and is developed with the help of the source. Figure 12.11 illustrates the methods of DLLDM and MAM toward the transmission of data packets with various attacker numbers. The packet distribution ratio for the two networks is reduced due to the

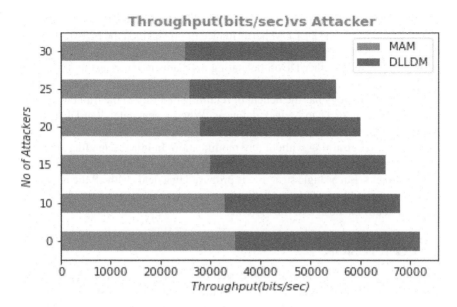

FIGURE 12.10   Throughput versus Attackers.

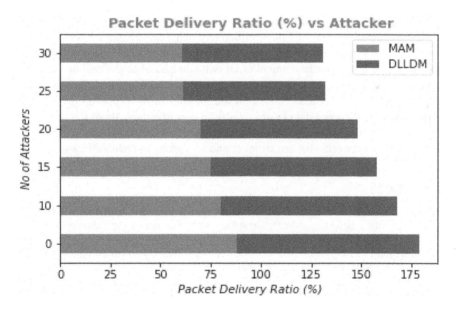

FIGURE 12.11   Packet Delivery Ratio versus Attackers.

increase in the number of attackers. As the network size surges, PDR declines rapidly. Since the cooperation of several malevolent nodes, the entire information-forwarding phase, the DLLDM shows a higher PDR than the MAM. The DLLDM generates with each node, in the routing process, with the faster outcomes, the data packet is destroyed or damaged. CHs are selected from the reliable environment to guarantee reliability. The DLLDM performance improved than the MAM, supplying the maximum number of packets of network data.

Compared with the other DDoS defending mechanisms, such as the MAM solution, the proposed DLLDM architecture is a great defending mechanism for DDoS attacks. This technology has the benefit of retaining mobility in the framework of the system during various DDoS attack defences. When a new defending DDoS attack process in a system is introduced or removed, the feature modules alone can be changed in system training without any modification. In other defending strategies, there are some drawbacks, such as flood-based techniques that do not defend against jamming attacks. But a single structure is proposed that deals with all the issues. The structure suggested benefits from the flawless interpretability of the attack reduction and is distinct of the design of the system.

12.7 CONCLUSION

In this chapter, with the help of the suggested DLLDM frame structure, the network is protected against DDoS attacks. For versatile and energy-effective cluster-based countermeasures, this robust detection mechanism is used along with LSTM-based DL for effective detection and separation purposes. Due to the decrease in energy consumption, the life span of the nodes is improved. The experimental result and the performance review of the method to DLLDM and MAM describe that the suggested DLLDM frame structure will most effectively achieve satisfactory results on the detection of DDoS attacks. For nodes with low to no mobility, this method is applicable. In the future, the enhancement of DL with an LSTM model will be addressed through the implementation of different optimization methods and the public data sets that contain these attacks. Based on the findings of the proposed model, more study of four-dimensional data to discover measures to remove these unknown multiple threats from a WSN is needed. The proposed framework results in creation of wireless sensor networks that are stronger and less prone to DDoS attacks.

REFERENCES

1. Kolandaisamy, R. et al. (2020). Adapted Stream Region for Packet Marking Based on DDoS Attack Detection in Vehicular Ad hoc Networks. *Journal of Supercomputing*, 76, 5948–5970, doi:10.1007/s11227-019-03088-x.
2. Bala, P. M., S. Usharani, and M. Aswin. (2020). IDS Based Fake Content Detection on Social Network Using Bloom Filtering. *2020 International Conference on System, Computation, Automation and Networking (ICSCAN), Pondicherry, India*, 1–6, doi:10.1109/ICSCAN49426.2020.9262360.
3. Adimoolam, M., A. John, N. M. Balamurugan, and T. Ananth Kumar. (2020). Green ICT Communication, Networking and Data Processing. In *Green Computing in Smart Cities: Simulation and Techniques*. Cham: Springer, 95–124.
4. Zhang, L., F. Restuccia, T. Melodia, and S. M. Pudlewski. (2019). Taming Cross-Layer Attacks in Wireless Networks: A Bayesian Learning Approach. *IEEE Transactions on Mobile Computing*, 18(7), 1688–1702, 1 July, doi:10.1109/TMC.2018.2864155.

5. Wang, Renqiang, Donglou Li, and Keyin Miao. (2020). Optimized Radial Basis Function Neural Network Based Intelligent Control Algorithm of Unmanned Surface Vehicles. *Journal of Marine Science and Engineering*, 8, 210, doi:10.3390/jmse8030210.

6. Agrawal, N., and S. Tapaswi. (2019). Defense Mechanisms Against DDoS Attacks in a Cloud Computing Environment: State-of-the-Art and Research Challenges. *IEEE Communications Surveys & Tutorials*, 21(4), 3769–3795, fourth quarter, doi:10.1109/COMST.2019.2934468.

7. Vishvaksenan, Kuttathati Srinivasan, and R. Rajmohan. (2019). Performance Analysis of Multi-Carrier IDMA System for Co-operative Networks. *Cluster Computing*, 22(4), 7695–7703.

8. Premkumar, M., and T. V. P. Sundararajan. (2020). DLDM: Deep Learning-Based Defense Mechanism for Denial of Service Attacks in Wireless Sensor Networks. *Microprocessors and Microsystems*, 79, 103278.

9. Kalaipriya, R., S. Devadharshini, R. Rajmohan, M. Pavithra, and T. Ananthkumar. (2020). Certain Investigations on Leveraging Blockchain Technology for Developing Electronic Health Records. *2020 International Conference on System, Computation, Automation and Networking (ICSCAN), Pondicherry, India*, 1–5, doi:10.1109/ICSCAN49426.2020.9262391.

10. Alshawi, Amany, Pratik Satam, Firas Almoualem, and Salim Hariri. (2020). Effective Wireless Communication Architecture for Resisting Jamming Attacks. *IEEE Access*, 8, 176691–176703.

11. Deiva Ragavi, M., and S. Usharani. (2014). Social Data Analysis for Predicting Next Event. *International Conference on Information Communication and Embedded Systems (ICICES2014), Chennai*, 1–5, doi:10.1109/ICICES.2014.7033935.

12. Marti, Sergio, Thomas J. Giuli, Kevin Lai, and Mary Baker. (2000). Mitigating Routing Misbehavior in Mobile Ad hoc Networks. *6th ACM Annual International Conference on Mobile Computing and Networking*, 255–265, August.

13. Thottam Parameswaran, Ambili, Mohammad Iftekhar Husain, and Shambhu Upadhyaya. (2009). Is RSSI a Reliable Parameter in Sensor Localization Algorithms: An Experimental Study. *IEEE Field Failure Data Analysis Workshop (F2DA09)*, 5, September.

14. Prakash, Kolla Bhanu. (ed). (2020). *Internet of Things: From the Foundations to the Latest Frontiers in Research*. Berlin: Walter de Gruyter GmbH & Co KG.

15. Gopalakrishnan, A., P. Manju Bala, and T. Ananth Kumar. (2020). An Advanced Bio-Inspired Shortest Path Routing Algorithm for SDN Controller Over VANET. *2020 International Conference on System, Computation, Automation and Networking (ICSCAN), Pondicherry, India*, 1–5, doi:10.1109/ICSCAN49426.2020.9262276.

16. Bala, P. M., and S. Hemamalini. (201). Efficient Query Processing with Logical Indexing for Spatial and Temporal Data in Geospatial Environment. *2019 IEEE International Conference on System, Computation, Automation and Networking (ICSCAN), Pondicherry, India*, 1–6, doi: 10.1109/ICSCAN.2019.8878743.

17. Sasikala, I., M. Ganesan, and A. John. (2018). Uncertain Data Prediction on Dynamic Road Network. *International Conference on Information Communication and Embedded Systems (ICICES 2014)*, 1–4, doi:10.1109/ICICES.2014.7033972.

18. Adimoolam, M., M. Sugumaran, and R. S. Rajesh. (2018). A Novel Efficient Clustering and Secure Data Transmission Model for Spatiotemporal Data in WSN. *International Journal of Pure and Applied Mathematics*, 118(8), 117–125.

19. Sundareswaran, P., R. S. Rajesh, and K. N. Vardharajulu. (2018). EGEC: An Energy Efficient Exponentially Generated Clustering Mechanism for Reactive Wireless Sensor Networks. *International Journal of Wireless and Microwave Technologies*, 8.

20. Rajesh, G., C. Vamsi Krishna, B. Christopher Selvaraj, S. Roshan Karthik, and Arun Kumar Sangaiah. (2018). Energy Optimised Cryptography for Low Power Devices in Internet of Things. *International Journal of High Performance Systems Architecture*, 8(3), 139–145.

21. Rajesh, G., X. Mercilin Raajini, and B. Vinayagasundaram. (2016). Fuzzy Trust-Based Aggregator Sensor Node Election In Internet of Things. *International Journal of Internet Protocol Technology*, 9(2–3), 151–160.

22. Jayapriya, Kalyanakumar, N. Ani Brown Mary, and R. S. Rajesh. (2016). Cloud Service Recommendation Based on a Correlated QoS Ranking Prediction. *Journal of Network and Systems Management*, 24(4), 916–943.

23. Kumar, T. A., A. John, and C. R. Kumar. (2020). 2 IoT Technology and Applications. *Internet of Things*, 43.

24. Selvi, S. Arunmozhi, R. S. Rajesh, and M. Angelina Thanga Ajisha. (2019). An Efficient Communication Scheme for Wi-Li-Fi Network Framework. *2019 Third International conference on I-SMAC (IoT in Social, Mobile, Analytics and Cloud) (I-SMAC)*, 697–701.

13 A Knowledge Investigation Framework for Crowdsourcing Analysis for e-Commerce Networks

Harsh Jigneshkumar Patel, Dipanshi Digga, and Jai Prakash Verma

CONTENTS

13.1 INTRODUCTION

Crowdsourcing analysis for e-commerce networks is the technology used to extract the review text data using textual analysis, including computer language and natural language processing methodologies (Jeonghee Yi, (13)) (Rosander, (16)) (Zornitsa, (24), 524–533). Opinions consist of six domains of work: extraction of opinion, sentiment analysis, analysis of subjection, sentiment mining, mining of reviews and affect, or emotion analysis. This technology is used to analyze people's sentiments, attitudes, appraisals, and emotions related to a particular product, service, and organization using an e-commerce network. The methods can be classified into two ways: a lexicon or a rules-based approach or a machine learning approach. The lexicon-based approach uses the sentiment dictionary to calculate the sentiment score of the word embeddings in the text sequence generated by the tokenizer (Shaiful, (19)), (Y Chen, (27)), (R. Pervaiz, (31)), (Huang, (34)), (Eapen, (36)). The machine learning approach is segmented into three methodologies: supervised, semi-supervised, and unsupervised learning (Jyoti, (3), 55–64), (Tjahyanto, (8), 38–45), (Q. Huang, (17)). Supervised machine learning algorithms focus on the concept of training with examples by improvising the parameters on each iteration after a training data point is fed in as input (Yuanbin, (7)), (V. N. Garla, (29), 992–998), (A. Khadka, (33)). The lexicon-based approach consists of a corpus-based approach and the dictionary-based approach. The machine learning method consists of a feature vector construction like the bag-of-words approach. The text data extracted is then analyzed by the machine learning algorithms for different purposes, be it sentiment mining or text summarization (Kamal, (6), 358–365), (Monisha, (12)), (Aravind, (22)).

By this proposed research work, the analysis is made simpler as the analysts would not have to work and analyze each of the reviews or text data in general. The representative data points would provide an overview of the situation or the problems in the product development phase in a particular company (Yuanbin, (7)), (A Khan, (30)). This work would also help in better and faster analysis of the foreign markets for strategists as the data required to analyze the foreign markets would be extensively large, which would make the analysis procedure computation- and data-intensive (Vinodhini, (5), 311–319) (Rosander, (16)). It would be beneficial to the industrial development process as well in consideration of Industry 4.0. Business analysts working in diverse industrial sectors could benefit from the process as the aspects or domain-persisting information can be mined easily with the help of branched information present in the databases. An e-commerce network consists of data pertaining to various fields of work, and the analysis of information gets tedious with increasing amounts of data to be manually analyzed. The proposed research work tends to make the analysis easier by interconnecting each node of the e-commerce network on the basis of the aspects entailed in the word corpus. Sociologists could benefit from the proposed research work as they could analyze the problems faced by civilians living in a particular area. The problems faced by the people in an area are generally the same in their aspects of nature, which creates the problem of data redundancy, thus decreasing the efficiency of the analysis procedure (Kiran, (18)), (Aravind, (22)), (Wanxiang, (10), 2111–2124). The method would also help companies analyze the major features in which the product is lagging with the help of feature-based sentence presentation and clustering (Tjahyanto, (8), 38–45), (Ricardo Gacitua, (23)).

13.1.1 Needs Analysis

With the explosion of Web 2.0, there has been a surge in the networks and the amount of information being produced by the users. For example, there has been an upsurge in the review data being produced on online platforms such as Amazon, Twitter, and the like (Tjahyanto, (8), 38–45) (Mikalai Krapivin, (25), 102–111). The amount of data on the web has been increasing exponentially in comparison to the amount produced in the last decade. With open platforms, people are free to share their opinions and experiences regarding a situation or a product. Many authors have previously tried to tackle the problem of analyzing the data online. But the text data being produced cannot be compressed in the aspect of the content being written or based on the amount of the space being occupied by them (Wanxiang Che, (10), 2111, 2124), (Y Chen, (27), 265–272), (R. Pervaiz, (31)). The amount of information can be analyzed in an efficient way that would not compromise the quality or the amount of content being provided. Having a vast amount of text data to analyze is a tedious and data-intensive task to execute, which makes analyzing such huge data sets uncompelling.

The text data in the field of data analysis consists of movie reviews, online product reviews, and articles related to a particular issue in society or the domain (Andrea Kő, (15)), (Chen, (20)), (D Fišer, (32)). The data could be related to the quality of the content of the movie, the actors' performances, the defects in a particular product, or any problem faced by the civilians in a country. Having tools to analyze and infer the issues and aspects of the data being provided is crucial as the results would be the basis of the action plans that would be undertaken to solve the conflict. The analysis of the text data is based on the three levels: the word level, the sentence level, and the document level, respectively (S Ramaswamy, (26), 170–178) (David Faure, (21)), (Karmen, (28)). These three levels of analysis have their uses, of which the sentence- and document-level analysis of the data is used for the aspect-based clustering and a review representation of the whole corpus.

According to Forbes, there are 2.5 quintillion bytes of data being produced every day, which is mainly due to the development of the Internet of things (IoT). More than 3.7 billion humans are using mobiles today which is an increase of about 8% since 2016 (Shaiful, (19)), (S Ramaswamy, (26), 170–178), (VN Garla, (29), 992–998). The data creation every minute consists of about 500,000 photos being uploaded on Snapchat, 4.1 million users watching video content on YouTube, 450,000 tweets on Twitter, and 46,000 Instagram posts (Zhiqiang Ge, (11)), (J Rashid, (35), 6573–6588). The data being generated are increasing exponentially, with estimation by IBM about the total data produced to reach about 35 zettabytes by the end of 2020, which was merely 800,000 petabytes in 2000. The data also consist of 16 million text messages, 1 million Tinder swipes, 156 million emails being sent, and 150,000 Skype calls being made (A Khan, (30)) (see Figure 13.1). According to Arne Holst, the data produced annually have been increasing dramatically and are estimated to reach 175 zettabytes by 2025. These data would include data from various industries, ranging from production to entertainment (Tom Young, (14)).

13.1.2 Contribution

This research contributes to various domains of industry dealing with text data for their business or product development using an e-commerce network. This is achieved by analyzing the various aspects and features of the products or the services for a better understanding of the opinions. Grouping the reviews to a particular aspect and finding a representative node helps in reducing the computation time and complexity by analyzing a single review opinion per aspect. Calculating relativity

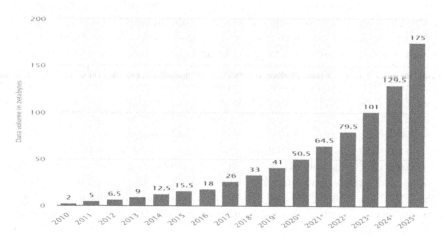

FIGURE 13.1   Data Production Annually.

indices between the nodes is implemented using the PageRank algorithm by creating transition matrices. The procedure is also tested successfully for various text data sets including the IMDb movie reviews data set and the Twitter database. This research contributes to various domains of industry dealing with text data for their business or product development using an e-commerce network. This is achieved by analyzing the various aspects and features of the products or the services for a better understanding of the opinions. Grouping the reviews to a particular aspect and finding a representative node helps in reducing the computation time and complexity by analyzing a single review opinion per aspect. Calculating relativity indices between the nodes is implemented using the PageRank algorithm by creation of transition matrices. The procedure is also tested successfully for various text data sets, including the IMDb movie reviews data set and the Twitter data set.

13.1.3 Chapter Organization

The rest of the chapter is organized as follows: Section 13.2 shows a comparative study of the related literature on the different research done. Section 13.3 presents the proposed system for the proposed research work presented. Section 13.4 presents the methodology and concepts applied to achieve the defined objectives. Sections 13.5 and 13.6 present performance analyses by describing the experiments and the experimental results. Section 13.7 discusses the conclusions and future work.

13.2   RELATED WORK

The research work of many researchers working in the same field is presented along with the methodology and the approaches of their sentence representation techniques. Paramita Ray ((1)) stated a mixed approach of deep learning and a rules-based method to improve aspect-based sentiment analysis in which he used both the working methodologies to learn about the aspects of the review data. The method uses a rules-based approach along with a seven-layer-deep convolutional network model for aspect-based extraction that includes different layers, that is, convolution, pooling, dropout regularization, full connection, and softmax activation layer. Edison ((2)) devised a novel deterministic approach for aspect-based opinion mining in tourism product reviews using the lexicon methodology with a rules-based structure. The sentiment orientation of the extracted expressions is done using the word rules, negation rules, and rules that assign a value to the words according to their orientations as 1, 0, and −1. Deshmuikh ((3), 55–64) proposed research work for building an entropy classifier for cross-domain opinion mining using semi-supervised learning. Transfer learning is used for analyzing the problem of a particular domain based on the learning obtained from the source domain. N-grams are used for capturing the target words and the context of the feature and the word used.

W. M. Wang ((4), 149–162) proposed a method for building an opinion mining technique for product reviews by summarizing the product features and the reviews related to it using Kansei engineering. The selection of the first few Kansei words is done manually and the Kansei words are divided into two different groups, which comprise the two orientations of the Kansei attributes then fed into the WordNet for the extraction of their synonyms and antonyms. Vinodhini ((5), 311–319) worked on a methodology to analyze drug satisfaction using supervised learning methods. Construction of the vector space representation of the following drugs is also implemented. The training data set is processed using the different machine learning algorithms and neural networks, namely, Support Vector Machines (SVMs), neural networks using PNNs (probabilistic neural networks), and neural networks using the radial basis function network, predicting the class of the drugs using the different models.

Kamal ((6), 358–365) devised a methodology to mine opinions for competitive intelligence of companies in a particular industry. The feature extraction from the text data is machine learning-based that is derived from the probability of the occurrence of the nouns or the noun phrases in the text-based on their frequency. The machine learning approach includes algorithms such as Naïve Bayes models, SVMs, artificial neural networks (ANNs), and maximum entropy classifiers. The lexicon-based approach uses the SentiwordNet lexicon, which consists of the sentiment score of each word with their corresponding sentiment score. The method uses the graphical relations between the nodes, be it synonyms, antonyms, hypernyms, or hyponyms. The hybrid approach uses machine learning and the lexicon-based approach by the implementation of SVM and particle swarm optimization, which is based on different frameworks of machine learning and principal component analysis. Yuanbin ((7)) researched a novel approach of opinion mining for review representation using a graph-based data analysis procedure. This includes the conversion of features into feature expressions and relating the expressions and the modifiers. This also includes individual opinion relation representation. Tjahyanto ((8), 38–45) proposed a method to utilize filters on object-based opinion mining for tourism reviews. The classification of the opinions is done using a Naïve Bayes model, an SVM, and a random forest model. The data filter is applied, which consists of 556 components of hotels and restaurants. Then the process consists of object extraction and determination of object orientation. Dau ((9), 1279–1292) proposed a method for a recommendation

TABLE 13.1

Comparative Analysis of Work Done by Different Researchers in the Area of Opinion Mining

Reference	Objective	Methodology	Approach	Pros	Cons	Accuracy
Ray, (1)	Improve aspect-based opinion mining.	PCA, hierarchical clustering, CNN, skip-gram	Deep learning and rules-based	High accuracy summarization and classification	The computation cost is very high.	87.12%
Edison ((2) 2014)	Improving tourism-based opinion mining	Aspect words aggregation rules Aspect aggregation rule Position aggregation rule	Rules and lexicon-based	Creation of data set for features related to the tourism domain	Poor aspect extraction	91.73%
Deshmuikh ((3), 55–64)	Efficient cross-domain sentiment analysis	SVM, Bag of words, N-grams classifier, Naive Bayes	Lexicon-based Deep learning and graph-based data analysis	Better accuracy in comparison to Naive Bayes and SVM	Accuracy falls when domains are not related	85.03%
Wang ((4) 201 8, 149–162)	Better summarization through Kanski attributes	Clustering rules using Kansei features and words	Unsupervised learning using Kansei engineering	Identification of opinions of most importance by comparison	Precision performance in critical reviews is low	90.39%
Vinodhini ((5), 2017, 311–319)	Analysis of drug satisfaction	PNN and RBFN	Deep learning–based	RBFN uses the localized radial basis function.	Gets trapped in an undesirable local minimum	86.38%
Kamal ((6) 2015, 358–365)	Opinion mining in competitive intelligence	ANN, SVM, Maximum Entropy, particle swarm optimization, PCA	Deep learning– and lexicon-based	Higher accuracy of sentiment classification	Does not cover all the aspects due to the absence of features in lexicons	87.91%

Reference	Approach	Methods	Category	Advantages	Limitations	Accuracy
(Yuanbin ((7))	Graph-based opinion representation	Standard sequential labeling method, ILP Formulas and minimum spanning tree	Graph-based opinion mining	Gives more aspect correlations than the SVM classifier	Confusion between the target relations and coordination labels	92.46%
Tjahyanto ((8), 38–45)	Filter-based opinion mining in the tourism domain	Naïve Bayes, SVM, and random forest classifiers with data filters	Machine learning and filter-based data mining	Increases the accuracy in feature extraction and aspect-based summarization process	It is constrained to a particular set of aspects and domains	90.27%
Dau ((9), 1279–1292)	Recommendation system exploiting aspects using deep learning	MCNN, tensor factorization, and SVM classifier	Deep learning and a rules-based system	Various aspects of a single opinion are analyzed.	The computation time and cost are comparatively higher than deep learning models.	91.52%

system exploiting aspect-based opinion mining with a deep learning method. The procedure included a MCNN (multichannel deep convolutional neural network) for the generation of aspect-specific rating of the product reviews. Then the aspect-based ratings are integrated into a tensor factorization (TF) for rating prediction. Table 13.1 depicts the various methodologies and approaches used in the related works for a better understanding of the field of opinion mining.

13.3 PROPOSED RESEARCH WORK

This chapter aims to propose a novel opinion mining architecture for the opinion review representation system using deep learning methodologies and a PageRank algorithm (refer to Figure 13.2) for an e-commerce network. The deep learning technology used for aspect mining from the text corpus by initially preprocessing the text to remove the noise from the data. The three deep neural network models are tested for finding the best performing model for aspect mining from the data which is followed by the PageRank algorithm being implemented on the mined aspects and the review nodes. Text preprocessing is done to decrease the noise present in the data that may lead to inconsistencies and reduced performance of the proposed system. The text data is preprocessed by the various functions from the Natural Language Toolkit (NLTK) library, which includes the removal of stop words, stemming and lemmatization, the removal of numbers and punctuations, and the tokenization of the words and conversion to sequences. Tokenization of the terms in the text corpus is done to decrease the computation complexity by analyzing each character of the text keywords and assigning a numerical value to the term instead. The tokens are arranged in the form of sequences for further processing of the data starting from preprocessing of the text to processing with deep learning technologies. The tokenization is done using the GLove word embedding, which is used to map the words to a particular positive integer. If certain terms of the text corpus are not present in the embedding, they are assigned a value separately to avoid inconsistency in the processes ahead. Stop words consist of parts of speech such as articles, prepositions, and conjunctions. They tend to increase the computational load of the system as they do not carry any semantic significance. Stemming and lemmatization are methodologies used for discarding the semantically redundant words in the sequence to reduce the computation cost. Simply stripping of suffixes is done with the help of stemming, but the resultant words do not make sense at times, whereas the lemmatization of the word sequences is done to generate words with semantic meanings. This is executed with the help of the NLTK library and the Porter Stemmer toolkits. Dependency parsing is done on the text data to relate the "main" words and the words that build a relationship between the keywords in the opinionated sentences.

There are different activation functions used in the LSTM neural network by different researchers, but the most accurate and efficient activation function known for the hidden layer is the ReLU, rectified linear activation, function whereas the sigmoid activation function is used in the last layer of the neural network. This is mainly because the sigmoid function is mainly used for discrete class classification, which hinders the performance of the overall neural network, whereas the ReLU is a continuous activation unit; thus, it is efficient to use for the computation of continuous

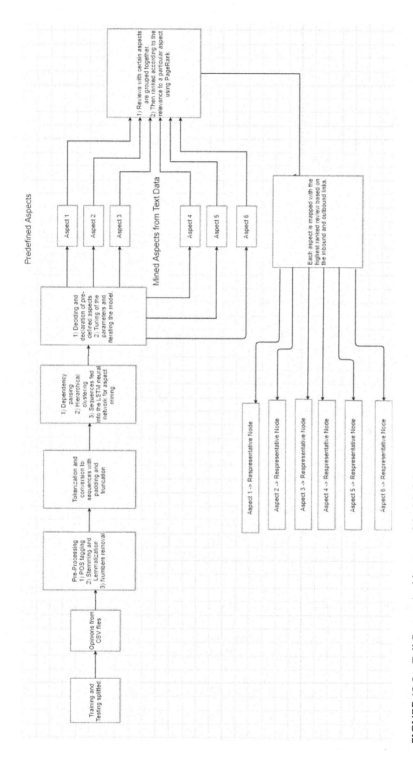

FIGURE 13.2 Full System Architecture of the Model.

values. The optimizer for the neural network is the Adam optimizer as it is known to have a better performance than the other optimizers used in the neural networks. To avoid overfitting and underfitting conditions in the results and training process, L2 and dropout regularization functions are used. The classification of the reviews according to the aspects is done on various levels of granularity, namely, the document level, the sentence level, and the word level. The word embeddings of the text data are created using the GLove corpus of word embeddings.

The context clarity of the reviews decreases from the document level through the sentence level to the word level, whereas the specificity of the review and analysis of the keywords for the aspect classification increases along the same hierarchy. At the sentence level of the reviews data, LSTM is used for the classification whereas, on the word level; the words are matched according to the aspect labels according to the similarity of the word vectors in the word embedding. This helps in the accurate aspect analysis and classification based on the context and the specific words affecting the same. Hierarchical clustering is used for the reduction of the errors of aspect extraction and mining of the aspects of the opinionated sentence using a set of predefined aspects present. After the classification of each review according to the aspect present, the reviews are clustered aspect-wise, and a representative review datapoint is selected based on the similarity that it holds with the rest of the reviews. This is achieved with the help of the PageRank algorithm by considering the reviews as the nodes of a graph and calculating the number of links between the reviews to identify the representative node according to the importance in the form of aspect that it holds. Markov's chain rule is used for the PageRank algorithm in which each review data point is considered as a state in the graph. A feature common to two review data points is considered the transition from one state to another with a probability.

A transition matrix is constructed for all the review data points and their transitions from one state to another with the help of transitions is represented as 1 in the matrix, whereas no transition between any two states is represented by 0. A stochastic matrix construction is targeted which consists of nonzero values in each row of the matrix and the values adding to 1. But there may be data points that may not be related to any of the other review data points by any aspect and end up as dangling states that are solved by adding one extra state having outgoing transitions to every state in the graphical cluster. Then the probability distribution of each of the data points is calculated for the calculation of the importance and similarity of the pages within the cluster. The links between the states of the graph are compared to find the one with the greatest number of links. The PageRank algorithm states that the state with the greatest number of links is considered to be the most important state. The corresponding scenario here means that the review data point having the most relevant aspects mentioned is chosen as the representative sentence.

13.4 METHODOLOGY, CONCEPT, AND TECHNOLOGY

The listed approaches of deep learning methodology are used for the various use cases of text processing and are analyzed and compared based on the evaluation metrics (Figure 13.3).

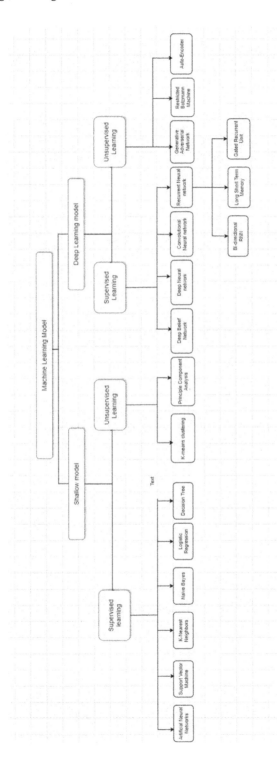

FIGURE 13.3 Machine Learning Taxonomy.

Feature Selection: The two types of features are crucial to be present in the feature vector, the first one being the explicit information being presented in the form of words in the text corpus and the second being the information extracted from the structural representation of each word token present in the sentence and the part of speech it plays. Embedding vectors that are pretrained are used for the classification of the words as important features as well. An important and crucial feature vector that is used is the dependency of the words on each other in the formation of semantic logic in the sentence. The parts of speech tagging is used for the execution as each part of speech is required to structure the information present in the raw form. Along with it is the problem of variable lengths of the input vectors, which is resolved by padding of vectors.

LSTM: The human thinking process never starts from the base level every time a person tries to solve a problem or thinks about situations, whereas machines are designed in a way that they have to be made to learn about the work or have to have information feed into them every time a particular problem is to be solved. This contrasts the nature of the neural networks, which states that the neural networks tend to function in the same way as the human brain. The recurrent neural networks tend to function in a way that helps them save information or the results that have been fed or computed by the activation functions at every node in the neural network. The networks are designed to feed the output of one activation unit to the other node ahead along with the information being inputted independently at each node. This helps the units learn the information from the past and remember the results in a contiguous manner. LSTM is a special kind of recurrent neural network (RNN) capable of learning long-term dependencies and is designed in the form of a chain-type structure being connected by the results fed from the previous layer of the neural network. The LSTM regulates the information being fed in the network with the help of various gates that are composed of a sigmoid neural net layer and a pointwise multiplication operation. Another version of LSTM is the LSTM GRUs, gated recurrent units. The GRUs are used to combine the input and output gates to the update gate, and the reset gate is the same as the forget gate.

CNN (convolutional neural network) is the deep neural network model mainly used for image processing that is executed with the help of convolutions of the data point. The image processing is done with filters being applied to the image pixels, or it may be the word embeddings in the text data. This helps with analyzing and efficiently processing the text data. Previously, the processing of text data was mainly done with the help of an RNN, but recently, the use of CNNs in the field of natural language processing has been increasing due to the RNNs being slow in the computation and processing of text data. Pretrained models, along with a faster softmax function, are used for the architecture of the CNN model. The major reason the RNN models were dropped was due to the text embeddings being processed in sequential flow, whereas a CNN model was used to process the text parallelly at the same time. CNN takes the embedding of text data and uses multiple layers of gated convolutions, with each layer consisting of a convolutional block that has two separate convolution outputs and one gating block which uses one convolutional output to gate the other. The normal convolutional layer has a window width of k, which processes the data word embeddings one at a time and the convolutional width overlapping with each other by one unit at a time. The activation functions generally used are the ReLU and the sigmoid functions. The real functions are generally used in the hidden layers as the values are more diverse than the output values of the sigmoid function and are thus used for more accurate results, whereas the sigmoid function is used for the classification of the data based on the output results to classify in either class.

The RNN model is based on the concept that the output of the previous layer is fed into the next layer, and the result that works as a part of the input that helps in remembering the results from various layers. The activation results consist of the input from the previous layers and an input particular to a node in the layer. This allows the network to learn the long-term dependencies in the text data being processed that, in turn, helps with aspect-based sentiment analysis and abstract text summarization, which results in a better understanding of the meanings of the text data by providing the context of the text. The structure consists of the three gates: the input, the output, and the forget gate. The process consists of the conversion of the text data into sequences using an embedding matrix

PageRank Algorithm: The aspect mining from the review text data is implemented using deep learning technologies, whereas the review representation system is implemented using the PageRank algorithm. This is the algorithm used by Google for the Google search engine by analyzing the importance of the web pages by calculating the measures needed for them. This is mainly done by calculating the importance metric by counting the number of links to and from a webpage and the number of visitors visiting the page. Weights are assigned to the links between the states of the algorithm. The states of the graph have a probability distribution of how likely a user would click on the link. The probability value in the distribution varies between 0 and 1. PageRank algorithm, in general, can be used for both directed and undirected graphs.

For example, consider a graph consisting of four web pages: W, X, Y, and Z. Transitions between the pages are considered as links between the states. Multiple links from one state to another is considered as a single link. The initial value of each page in the graph is assumed to be 0.25. If W, X, and Y are the links to the state Z, then the total value transferred to page Z is 0.75.

$$PR(Z) = PR(W) + PR(X) + PR(Y). \qquad (13.1)$$

Damping factor for states in the graph: The PageRank algorithm assumes a user who randomly clicks on the links on a web page. The probability of the user to continue clicking randomly on the pages is called the damping factor. For the final value of the web pages, the damping value factor is subtracted from 1, and the resultant term is added to the product of the damping factor and the value of the sum of the incoming PageRank values. Here, the review opinions are considered as nodes in the graph and the relativity indices between the nodes are computed to compare the relativity and the aspects mined from the review nodes. There are several models based on the PageRank algorithm according to the different situations that occur during the surfing of web pages by random users.

13.5 IMPLEMENTATION AND EXECUTION

13.5.1 Execution Environment and Data Set Selection

The entire model building and execution are done using Jupyter Notebooks with Anaconda, which provides a runtime environment for the python machine learn ing and deep learning models to be executed. All the processing of the data is done using the graphic processing units (GPUs) present for data-intensive work jobs. The Amazon reviews data set is used for the analysis that is mapped with the corresponding IDs of the users. The data set consists of nine columns. The Twitter text data set consists of 1.6 million tweets from various domains. The text data set is extracted using the Twitter API. The tweets consist of multiple aspects ranging from technical to social purposes.

The various fields of the data set consist of the following: Polarity: This label is used to mark the sensitivity of the tweet analyzed with the help of sentiment analysis algorithms.

1. IDs: This represents the IDs of the tweets being analyzed by the deep learning models or machine learning algorithms.
2. Date: This field mentions the date of posting of the tweet on a Twitter handle. Flag: This field represents any query related to the tweet if present. The absence of a query is denoted by the value NO_QUERY.
3. User: The label represents the username of the user who posted a particular tweet on the internet.
4. Text: This contains the text data to be analyzed for the various insights of the data.
5. IMDb Data Set of Movie Reviews: The data set consists of 50,000 movie reviews in text form which is mainly used for sentiment analysis but here the review text corpus is used to mine the aspects from the given text and choose the representative review node for a particular aspect mined by the LSTM network.

The data set consists of the following fields:

1. Text Data: This is the data fed in to mine the aspects from the text and are further used for finding the representative review node.
2. Label: The labels represent the sentiment of the review text written for a particular movie. The two values for this field are positive and negative.

13.5.2 Data Preprocessing

Data preprocessing is needed before processing of the text data, which causes noise in the entire analysis procedure. This may include redundancies, numerical data of no significance as the analysis does not include statistics, lowercasing of the text to avoid confusion, and the removal of punctuation. Stop words like *"is"*, *"the"*, and *"and"* do not carry any semantic significance and just increase the computation load for the processor and removal of the stop words would let the analyzer focus on the more important words in the text corpus the noise from the text data, which includes many of the special characters decreases the accuracy of the procedure. Parts-of-speech tagging is done to categorize the text data into the respective parts of speech to help analyze the useful words among the entire corpus. This helps the process get more granular information about the text data, be it for sentiment analysis, text summarization, or aspect extraction. Most of the preprocessing tasks are implemented with the help of the NLTK library. The normalization of the text includes stemming and lemmatization of the data, which means tracing the words back to their root words so that redundancies are reduced.

After the comparative analysis of the results and the preprocessing techniques used by the researchers in the research work (Table 13.2), the list of the steps taken for the preprocessing of the text data is as follows: (a) Remove HTML tags. (b) Remove whitespaces. (c) Lowercasing of all the characters. (d) Removal of numbers. (e) Removal of stop words. (f) Lemmatization. (g) Parts-of-speech tagging. This set of preprocessing techniques, when applied together, yielded a higher performance of aspect mining as well as better feature extraction procedure with noise removal from the text corpus.

13.5.3 CHARACTERISTICS AND LIMITATIONS OF THE DATA SET

The Amazon review data set is easily available on Kaggle. The data set contains the product reviews and the metadata from Amazon from May 1996 to July 2014. The number of text reviews in the entire data set is 142 million. The data were collected in the form of JSON formats. The two types of data sets needed include a metadata file that includes all the information about the products, while the other one is used to contain the text review data generated by the users. The JSON files are converted into. csv format for processing in python using pandas, which makes it more comfortable to work with machine learning and deep learning models. The fields of the data in the Amazon review data set are (1) ReviewerID, a unique ID generated by Amazon to each customer; (2) as in, the product ID of the product; (3) reviewerName, the name of the reviewer; (4) Helpful, the rating of helpfulness of the product to solve problems; (5) reviewText, the text that has to be processed; (6) overall, overall rating given by any user of the product; (7) Summary, a summary of the review; (8) unixReviewTime, review time (unix time); and (9) reviewTime, review time. There are many limitations to the usage of the Amazon data set amongst which the short-length limitation affects the most as the reviews cannot be written and explained in a more brief and concise way. The data set does not understand the difference between the words when proper nouns and normal usage may seem to be the same and not easily differentiable. The data set contains the corresponding reviewID, which makes it easy to map the text data back to its owner to ease the communication process.

13.5.4 EXECUTION AND IMPLEMENTATION

Data visualization was done using Tableau. It helps in visualizing humongous amounts of data in a very efficient and interpretable way. The data, being in JSON format, could

TABLE 13.2

Preprocessing Techniques Used by Various Researchers in the Corresponding Related Work

Preprocessing Techniques	Remove HTML Tags	Expand Contractions	Remove Whitespace	Lowercasing Alphabets	Removal of Numbers	Removal of Stop Words	Lemmatization	POSTagging
Paramita (1)	Yes	Yes	No	Yes	Yes	Yes	No	Yes
Edison (2)	Yes	Yes	Yes	Yes	No	Yes	Yes	Yes
Deshmuikh (3)	Yes	No	No	Yes	No	Yes	Yes	Yes
W.M. Wang (4)	No	Yes	Yes	Yes	Yes	Yes	Yes	Yes
Vinodhini (5)	Yes	No	Yes	Yes	Yes	Yes	Yes	Yes
Kamal (6)	No	Yes	Yes	Yes	No	No	No	Yes
Yuanbin (7)	No	Yes	Yes	Yes	Yes	Yes	Yes	Yes
Tjahyanto (8)	Yes	Yes	Yes	Yes	No	No	Yes	No
Dau (9)	Yes	No	No	Yes	Yes	No	Yes	Yes
Che (10)	Yes	No	Yes	No	Yes	Yes	Yes	No

not be processed and visualized using Tableau; thus, the data were stored in SQL. MySQL works well with JSON format of data, and the data cleaning was conducted using MySQL, which helps in erasing redundant and missing values from the storage, which helps maintain the consistency of the text data for processing and visualization. The data are visualized to analyze the number of reviews being written in a period and the increase in the number of reviews from one year to another. Different analyses based on the content are done using Tableau, be it the length of the sentence or the helpful votes received from the form. The. csv file of the opinionated sentences is loaded and visualized and analyzed for the useful aspects that can be gained for a few predefined aspects to group the opinionated sentences. Then the text data are preprocessed with the help of the tools mentioned which makes it ready for opinion mining.

The text length is analyzed for padding and truncating of the data, which is not to be considered to reduce the loss of the computation power used during the whole process of aspect mining and analysis. After the analysis and the preprocessing of the text is completed, the text is converted into sequences that are fed into the deep neural networks for the procedure of aspect mining. After the aspects are mined and the opinionated sentences are grouped according to their suitable aspects, the PageRank algorithm selects the representative review amongst the review nodes for a particular aspect. The use of data loaders and batching is done to split the data into batches for proportionate computing power consumption and does not lead to the overuse of the resources.

13.5.5 MODEL ARCHITECTURE

The model architecture consists of encoder–decoder bidirectional LSTM neural network, which mines the aspects present in the text corpus with the help of the feature vector being created using the encoder and the decoder modules. The encoder trains on the tokens for the building of a precise feature vector. The encoder is used for the creation of the word vector of the entire sentence mentioned and the further feeding of information extracted from the sentence being used as an input into the classifier. The decoder is used in the architecture as the classifier in which the input is the sequence of tokens, which are in the form of features being extracted by the initial process of feature extraction and, at the same time, the vector, which represents the entire sentence that is encoded by the encoder module of the architecture. Bidirectional LSTM is used for recovering with the problem of vanishing gradients, which was evident in the previously used RNNs. The use of LSTM is mainly due to the capturing of the dependencies between the tokens, which are converted to feature vectors with the help of the encoders. The use of the bidirectional LSTM instead of the unidirectional LSTM is due to the loss of dependency in the information if the input information is from one side only as the context of the reversed side is not considered to be effective. In the bidirectional LSTM, one layer of the network contains the input from the forward sentence word tokens whereas the backward layer takes input from the inverted sentence tokens in the form of feature vectors. The fusion of the both feature vectors in the layer of BiLSTM, the future and past information help create context for the features present in the feature vector.

The attention mechanism is used after the encoding of the tokens into feature vectors and selection of the terms with the most importance based on the information available from the past and future information and prediction of the token as a target word. The difference, which leads to the better prediction of the target words, is the fact that the context vector used for the sentence representation while encoding the tokens is not used for the attention mechanism. New context vectors are created for each token present with the help of the grammatical relationship between the words present in the sentence using parts of speech to uncover each connection between the words present in the sentence representation. Outputs of the encoder serve as inputs for the calculation of the weighted sum of the inputs, which results in the context vector, whereas the attention weights are calculated based on the dependencies between the target word and the sentence representations in the form of the vector present.

A comparison of the three methodologies of deep learning models was conducted by feeding Amazon reviews data set for mining the aspects from the text corpus, which was followed by the sentence representation system by PageRank algorithm, which works with the same efficiency for all three deep learning neural network models. The difference in the evaluation metrics for the process between the CNN and the LSTM model is the range of the long-term dependencies between the tokens in the text corpus. The LSTM network consists of gates, which play a crucial role in the dependencies as the forget and update gates regulate the information to be remembered during the course of the process of analysis of the text corpus.

Table 13.3 analyzes the different technologies used by the particular researchers for the projects mentioned in the related works. The technologies range from deep learning neural network models to lexicon-based methodologies. This also encounters the ensemble learning being used that is known to increase the efficiency and accuracy by applying multiple methodologies to the text corpus being analyzed. Research projects with higher accuracies use the LSTM networks for the analysis and aspect mining of the text corpus. Table 13.4 provides a comparative analysis of the technologies used in the related research works for aspect mining and clustering.

As mentioned, and highlighted in Figure 13.4, the deployment of the proposed research helps with representative node analysis and efficient business modeling in Industry 4.0 with the usage of aspect mining and node analysis on textual data present in the e-commerce network pertaining to a domain.

TABLE 13.3
Comparative Analysis of Various Deep Neural Networks

Fields of Comparison	CNN	RNN	LSTM
Accuracy in Aspect Mining of Predefined Aspects	78%	71%	87%
Overall Accuracy	81.723%	76.279%	91.0823%
Long-Term Dependencies	No long-term dependencies between texts.	Yes	Yes. Also contains Gated recurrent units to decrease the number of gates and increase the accuracy.
F1 Score	93.62%	94.84%	95.83%

TABLE 13.4
Technologies and Methodologies Used in the Relevant Works

Technology Used by the Related Works	Long Short-Term Memory	Convolution al Neural Networks	Recurrent Neural Networks	Graph-Based Analysis	Rules and Lexicon-Based
Paramita (1)	Yes	No	No	No	Yes
Edison (2)	No	No	No	No	Yes
Deshmuikh (3)	No	Yes	No	Yes	Yes
Wang (4)	No	No	No	No	Yes
Vinodhini (5)	No	No	Yes	No	No
Kamal (6)	Yes	No	Yes	No	Yes
Yuanbin (7)	No	No	No	No	Yes
Tjahyanto (8)	No	Yes	No	No	Yes
Dau (9)	No	No	Yes	No	No

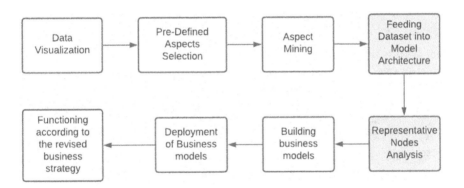

FIGURE 13.4 Deployment of Proposed Research Work in Business Modeling in Industry 4.0.

13.6 RESULTS AND DISCUSSION

The training accuracy of the proposed research deep neural network model is 91.0823% (Table 13.3). This result states that the efficient use of the representative nodes makes the job easier for the analyst as the representative review opinion considers all the review nodes being mentioned related to a particular aspect with almost all necessary information needed for the same. The predefined and mined aspects are used for testing the system based on the evaluation metrics.

This depicts the accuracy of PageRank algorithm for clustering and finding the representation review nodes for the various aspects mined by the deep neural network models and the predefined aspects (Figure 13.4). In Figure 13.5, the accuracy level of the model on the pre-mined aspects of the corpus increases up to five epochs and decreases after due to the overfitting of the neural network model resulting in

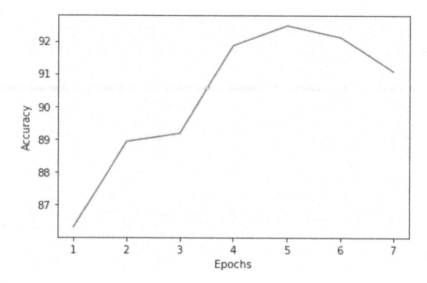

FIGURE. 13.5   Accuracy–Epoch Graph.

TABLE 13.5
Test Results on the Three Text Data Sets

Evaluation Metric	Amazon Reviews Data Set	Twitter Tweets Data Set	IMDb Movie Reviews Data Set
Accuracy	91.0823%	83.591%	90.830%
F1 Score	95.83%	82.019%	86.791%

inconsistencies in the accuracy in the epochs thereafter. This methodology of aspect-based review representation system works with more accuracy and F1 score for all the other data sets used for analysis. The comparative work for the data sets used CNNs and unidirectional RNNs for the aspect mining whereas in the proposed research work, LSTM was used for aspect mining. (Table 13.5)

13.7 CONCLUSION

This chapter presented a novel approach to solve the problem of redundancy and wastage of computational resources with the help of review nodes representation system. The review text data were mined for the various aspects and features present in the text corpus. This was done using the LSTM network, which produced a higher accuracy compared to the CNN and the RNN. Then the review text data were clustered on the basis of the aspects that the nodes consist of with the help of PageRank algorithm, which calculates a relativity index that helps the clustering process of the nodes and the selection of the representative node for each aspect mined from the text corpus. The accuracy of the aspect mining along with the review node representation was 91.0823%. The F-score, which considers both the precision and recall evaluation metrics for the calculation, also proves to be higher than the other neural networks with CNN producing 93.62%, while the RNN achieved a value of 94.84% and the LSTM neural network achieved a value of 95.83%. This proves an improvement in the opinion mining procedure that analysts have used in the past. The procedure was applied to different text data sets, like the Twitter data set and the IMDb review data set, and depicts the universality of the procedure being able to perform with all forms of text data. The advantages of this proposed architecture are numerous as it saves time and labor needed in the analysis of the information present in the e-commerce network, resulting in more functional procedures of Industry 4.0. The application of this procedure is appropriate for various sectors, as mentioned in the chapter, and in financial sectors, where news articles play a crucial role along with the redundant information being provided to the analysts. In the research work, predefined aspects were also used for the analysis procedure. The future work could consist of the mined aspects that could directly be analyzed according to the system procedure. Future work could also help with classifying the most important aspects that a team must work on for improvement with the help of sentiment analysis. The sentiment score related to each aspect can be compared to check the importance of each feature or aspect in the entire product.

REFERENCES

1. Ray, Paramita. (2018). A Mixed Approach of Deep Learning Method and Rule-Based Method to Improve Aspect Level Sentiment Analysis. *Applied Computing, and Informatics*, September, doi:10.1016/j.aci.2019.02.002.
2. Edison, M. T. (2014). A Novel Deterministic Approach for Aspect-Based Opinion Mining in Tourism Products Reviews. *Expert Systems with Applications*, 41, December.
3. Deshmuikh, Jyoti S. (2018). Entropy-Based Classifier for Cross-Domain Opinion Mining. *Applied Computing and Informatics*, 14(1), 55–64, January.
4. Wang, W. M. (2018). Extracting and Summarizing Affective Features and Responses from Online Product Descriptions and Reviews: A Kansei Text Mining Approach. *Engineering Applications of Artificial Intelligence*, 73, 149–162, August.
5. Vinodhini, G. (2017). Patient Opinion Mining to Analyze Drug Satisfaction Using Supervised Learning. *Journal of Applied Research and Technology*, 15(4), 311–319, August.
6. Kamal, A. (2015). Product Opinion Mining for Competitive Intelligence. *Procedia Computer Science*, 73, 358–365.
7. Yuanbin, W. (2011). Structural Opinion Mining for Graph-based Sentiment Representation. *Proceedings of the 2011 Conference on Empirical Methods in Natural Language Processing*, 1332–1341.
8. Tjahyanto, A. (2017). The Utilization of Filter on Object-Based Opinion Mining in Tourism Product Reviews. *Procedia Computer Science*, 124, 38–45.
9. Dau, Aminu. (2020). Recommendation System Exploiting Aspect-Based Opinion Mining with Deep Learning Method. *Information Sciences*, 512, 1279–1292, February.
10. Che, Wanxiang. (2015). Sentence Compression for Aspect-Based Sentiment Analysis. *IEEE/ACM Transactions on Audio, Speech, and Language Processing*, 23(12), 2111–2124, December.
11. Ge, Zhiqiang. (2017). Data Mining and Analytics in the Process Industry: The Role of Machine Learning. *IEEE Access*, 5, 20590–20616, 26 September.
12. Kanakaraj, Monisha. (2015). Performance Analysis of Ensemble Methods on Twitter Sentiment Analysis Using NLP Techniques. *Proceedings of the 2015 IEEE 9th International Conference on Semantic Computing (IEEE ICSC 2015)*, https://ieeexplore.ieee.org/document/7050801.

13. Yi, Jeonghee. (2003). Sentiment Analyzer: Extracting Sentiments About a Given Topic Using Natural Language Processing Techniques. *Third IEEE International Conference on Data Mining*, 22, November.

14. Young, Tom. (2018). *Recent Trends in Deep Learning Based Natural Language Processing*, arXiv:1708.02709v8 (cs.CL), 25 November, https://arxiv.org/pdf/1700.02709.pdf.

15. Kő, Andrea. (2010). *Research Challenges of ICT for Governance and Policy Modelling Domain—A Text Mining-Based Approach.* Springer Lecture Notes in Computer Science Book Series (LNCS). Heidelberg: Springer, vol. 9831.

16. Rosander, O. (2018). *Email Classification with Machine Learning and Word Embeddings for Improved Customer Support.* Sweden: Blekinge Institute of Technology.

17. Huang, Q. (2018). *CP-10 — Social Media Analytics, the Geographic Information Science & Technology Body of Knowledge*, first Quarter, John P. Wilson, https://scholar.google.com/.

18. Badwaik, Kiran. (2017). Towards Applying OCR and Semantic Web to Achieve Optimal Learning Experience. *2017 IEEE 13th International Symposium on Autonomous Decentralized System (ISADS)*, 8 June, https://www.secs.oakland.edu/~Mahmood/.

19. Bin Rodzman, Shaiful Bakhtiar. (2017). The Implementation of Fuzzy Logic Controller for Defining the Ranking Function on Malay Text Corpus. *2017 IEEE Conference on Big Data and Analytics*, https://ieeexplore.ieee.org/document/8284113.

20. Chen, Po-Hao. (2018). Integrating Natural Language Processing and Machine Learning Algorithms to Categorize Oncologic Response in Radiology Reports. *Journal of Digital Imaging*, 31(2), 178–184, April, doi:10.1007/s10278-017-0027-x.

21. Faure, David. (1999). Knowledge Acquisition of Predicate-Argument Structures from Technical Texts Using Machine Learning: The System Asium, Knowledge Acquisition, Modeling and Management. *11th European Workshop, EKAW'99, Dagstuhl Castle, Germany*, 329–334, 26–29 May.

22. Joshi, Aravind K. (2005). *Ranking and Reranking with Perceptron.* Berlin: Springer Science + Business Media, Inc., Manufactured in the Netherlands.

23. Gacitua, Ricardo. (2007). A Flexible Framework to Experiment with Ontology Learning Techniques. *Research and Development in Intelligent Systems XXIV, Proceedings of AI-2007, the Twenty-Seventh SGAI International Conference on Innovative Techniques and Applications of Artificial Intelligence.* Cambridge: SGAI, December.

24. Kozareva, Zornitsa. (2006). Paraphrase Identification on the Basis of Supervised Machine Learning Techniques. *Advances in Natural Language Processing*, 524–533.

25. Krapivin, Mikalai. (2010). Key Phrases Extraction from Scientific Documents: Improving Machine Learning Approaches with Natural Language Processing. *The Role of Digital Libraries in a Time of Global Change*, 102–111.

26. Ramaswamy, S. (2018). Customer Perception Analysis Using Deep Learning and NLP. *Procedia Computer Science*, 140, 170–178.

27. Chen, Y. (2011). Applying Active Learning to Assertion Classification of Concepts in Clinical Text. *Journal of Biomedical Informatics*, 45(2), 265–272.

28. Karmen, C. (2015). Screening Internet Forum Participants for Depression Symptoms by Assembling and Enhancing Multiple NLP Methods. *Computer Methods and Programs in Biomedicine*, 120(1).

29. Garla, V. N. (2012). Ontology-Guided Feature Engineering for Clinical Text Classification. *Journal of Biomedical Informatics*, 45(5), 992–998, October.

30. Khan, A. (2010). A Review of Machine Learning Algorithms for Text-Documents Classification. *Journal of Advances in Information Technology*, 1(1), February.

31. Pervaiz, R. (2020). A Methodology to Identify Topic of Video via N-Gram Approach. *IJCSNS International Journal of Computer Science and Network Security*, 20(1), January.

32. Fišer, D. (2017). Legal Framework, Dataset and Annotation Schema for Socially Unacceptable Online Discourse Practices in Slovene. *Proceedings of the First Workshop on Abusive Language Online*, January, https://1library.net/document/q7w0j8dz-framework-dataset-annotation-socially-unacceptable-discourse-practices-slovene.html.
33. Khadka, A. (2018). Using Citation-Context to Reduce Topic Drifting on Pure Citation-Based Recommendation. *Proceedings of the 12th ACM Conference on Recommender Systems*, 362–366, September.
34. Huang, W. (2015). A Neural Probabilistic Model for Context-Based Citation Recommendation. *Proceedings of the Twenty-Ninth AAAI Conference on Artificial Intelligence*, 2404–2410.
35. Rashid, J. (2019). A Novel Fuzzy k-Means Latent Semantic Analysis (FKLSA) Approach for Topic Modeling Over Medical and Health Text Corpora. *Journal of Intelligent & Fuzzy Systems*, 37(5), 6573–6588.
36. Eapen, B. R. (2020). LesionMap: A Method and Tool for the Semantic Annotation of Dermatological Lesions for Documentation and Machine Learning. *Innovations in Dermatological Electronic Health Records*, 3.

14 Intelligent Stackelberg Game Theory with Threshold-Based VM Allocation Strategy for Detecting Malicious Co-Resident Virtual Nodes in Cloud Computing Networks

Shawakat Akbar Almamun,
E. Balamurugan, Shahidul Hasan,
N.M. Saravana Kumar, and K. Sangeetha

CONTENTS

DOI: 10.1201/9781003107477-14

14.1 INTRODUCTION

Individual users and conglomerates are enabled to use software, platform and infrastructure as a service using the internet in cloud computing (CC). They do not need to buy, manage and develop their own (1). The operational cost of companies is minimized using CC while maximizing operational efficiency. There is growth in CC due to the emergence of the Internet of Things (IoT); about 6.3 billion devices are connected in 2003. The collection of data centres provides computing power to CC environments, and high-speed networks are used for interconnecting in various locations (2).

A distributed computers clusters form a cloud in CC. Over a network, remote users are provided with on-demand computational services or resources. The coordination of information technology resources is assisted using resource management mechanism corresponding to management actions done by cloud providers and consumers. It allocates resources to resource consumers from resource providers. The dynamic re-allocation of resources is allowed using this resource management, so the available capacity can be used effectively by the user (3).

Over the internet, the process of assigning available resources to required cloud applications in CC is termed as resource allocation (RA). According to predefined resource allocation policies, resources are allocated to competing requests by an Infrastructure as a Service (IaaS) cloud (4). Services are withheld by RA, if the allocation is not managed properly. So service providers are allowed to manage resources of every individual module to solve this problem. Resource management includes RA, and in an economic way, the available resources are assigned using this.

In today's computing, an attractive, as well as widely used, technology corresponds to virtualization (5). A highly optimized hardware utilization is enabled by the ability of sharing a single physical machine's resources between various isolated virtual machines (VMs). It leads to an ease of management as well as a migration of virtual system compared with physical counterpart. This leads to new computing paradigms and architectures (6). In CC, virtualization is a key component in specific. However, new security challenges are raised by the addition of another abstraction layer between software and hardware.

Widely used VM managers may not be assumed as a fully secured as shown by few security problems related to virtualization use. So this leads to important number of studies focusing on security enforcement in virtual systems due to the addition of this observation to ever-growing virtualized architectures popularity. Most basic guarantee provided by cloud service providers is privacy protection and information security in the cloud. So cloud security receives great attention (7).

Physical resources are virtualized using cloud platform and flexible storage and computing are provided by this. So various VM types can be used by a user.

Different users placed on the same host initiated the concept of co-resident VMs (or co-location VMs). The same physical resources, such as memory and the CPU, are shared by them.

A motivated and malicious adversary is considered by co-resident threat in CC that is not affiliated to a cloud provider. Legitimate cloud customers launching virtual servers in internet-facing instances for doing work for their businesses are termed as victims (8). The adversary, who is perhaps a business competitor, wishing to use novel capabilities provided to them by cloud co-residency for discovering relevant information about their target's business.

Compromising a victim's machine or reading private data is included in this. Subtler attacks like launching a denial-of-service (DOS) attack or load measurement on a victim's server are also included in this. Full freedom is given to the adversary for launching and controlling an arbitrary cloud instance. Masquerading is another legitimate cloud customer. For general use of a third-party cloud, it is necessary, and the cloud is a trusted component.

In co-residency recognition, however, virtualization side channels are a peril. Co-residency can be recognized utilizing a cross-VM secret channel as a ground truth. While further developed techniques for the fruitful arrangement are sketched out, for example, manhandling transient territory of occurrence propelling; it is demonstrated that a beast power approach is likewise unobtrusively effective (9).

Taking on the appearance of a real client, an aggressor can dispatch numerous cases, play out the co-residency check, end and rehash until the ideal arrangement is acquired. A few cross-VM data spillage assaults are additionally plotted, for example, heap profiling and keystroke timing assaults. Be that as it may, freely affirmed that a significant number of the methodologies, for example, the utilization of credulous system tests, are not, at this point, pertinent on the Amazon Elastic Computing Cloud (EC2).

This, joined with scholarly recommendations that better disconnect cross-VM impedance impacts, makes co-residency discovery altogether increasingly troublesome right now (10). So, for preventing this kind of attack, the deployment of VM with an effective, as well as secured, allocation strategy is focused in this work, which reduces VM co-residence probability.

With three optimization objectives, namely energy consumption, security and load balancing, proposed a threshold-based VM allocation strategy.

The remainder of the chapter is organized as follows: Section 14.2 reviews cloud security and co-resident attack detection techniques. Section 14.3 provides a system model, Stackelberg game theory concepts and threshold-based VM allocation strategy. Section 14.4 illustrates experimental analysis. Section 14.5 deals with the conclusions and future work.

14.2 LITERATURE SURVEY

In a cloud environment, recent attack detection mechanisms and RA methods that produce better results are described in this section.

Hasan et al. (11) used CPU scheduling with an optimized round-robin algorithm for implementing an enhanced technique to allocate cloud resources. Arrived process are distributed with a dynamic threshold mechanism (DTM), which arranges processes in an effective manner. To prevent any resource starvation, precise consideration of the properties of processes is done.

The time specified for every process for executing CPU computation in every iteration dynamically is termed dynamic time quantum. With respect to context switch, turnaround time and average waiting time, better performance is given by the proposed algorithm when compared with other algorithms.

Goel et al. (12) proposed a time-shared policy basically centred on tedious issues by utilizing scheduling calculations to get the fitting outcomes immediately. As the fundamental objective of the distributed computing utilization of disseminated assets and applying them to accomplish a higher throughput, execution and tackling large-scope figuring issues.

Bates et al. (13) introduced a co-occupant watermarking, a traffic examination assault that permits a vindictive co-inhabitant VM to infuse a watermark signature into the system stream of an objective occurrence. This watermark can be utilized to exhilarate and communicate co-residency information from the physical machine, trading off separation without dependence on interior side channels. Subsequently, this methodology is hard to shield without expensive underutilization of the physical machine.

Assess co-occupant watermarking under an enormous assortment of conditions, the framework burdens and the equipment arrangements from a neighbourhood lab condition to creation cloud situations. Furthermore, exhibit the capacity to start an undercover channel of 4 bits for each second and can affirm co-residency with an objective VM example in under 10 seconds. And furthermore, show that inactive burden estimation of the objective and ensuing conduct profiling is conceivable with this assault. This examination shows the requirement of a cautious plan for equipment to be utilized in the cloud.

Sundareswaran et al. (14) built a heap-based estimation attacks conceived explicitly because of the virtualization in cloud frameworks. Here, build up a structure to distinguish these assaults dependent on the perception that the occasions occurring during the assaults lead to a recognizable succession of exemptions. Also, test the precision of the system utilizing the Microsoft Azure framework.

Han et al. (15) proposed a guard system that makes it troublesome and costly for aggressors to accomplish co-home in any case. In particular, (1) distinguish the expected contrasts between the practices of aggressors and lawful clients; (2) apply bunching examination and semi-managed learning procedures to group clients; (3) model the issue as a two-player security game, and give a nitty gritty investigation of the ideal systems for the two players; (4) exhibit that the assailant's general expense is expanded drastically by one to two significant degrees because of the proposed safeguard component.

Zhang et al. (16) presented Home Alone, a framework that lets an occupant check their VMs' elite utilization of a physical machine. The key thought in Home Alone is to modify the typical use of side channels. As opposed to abusing a side channel as a vector of assault, Home Alone uses a side channel (in the L2 memory store) as a novel, cautious identification apparatus.

By breaking down the store utilization during periods in which "agreeable" VMs facilitate to stay away from bits of the reserve, an occupant utilizing Home Alone can recognize the action of a co-inhabitant "enemy" VM. Key specialized commitments of Home Alone incorporate grouping strategies to investigate reserve utilization and visitor working framework piece adjustments that limit the presentation effect of well-disposed VMs avoiding observed store divides. Home Alone requires no change of existing hypervisors and no unique activity or participation by the cloud supplier.

Abazari et al. (17) introduced a novel multi-target assault reaction framework. In this consider reaction cost, co-residency danger and VM connections to choose an ideal reaction in face of the assault. Ideal reaction determination as a multi-target enhancement issue ascertains elective reactions with the least danger and lowest cost. This technique gauges the danger level depending on the cooperation diagram and proposes appropriate countermeasures dependent on danger type with the lowest expense. Test results show that the framework can recommend ideal reactions depending on the present status of the cloud.

Beloglazov et al. (18) proposed a vitality mindful allotment heuristics arrangement server farm assets for customer applications in a manner that improves vitality effectiveness of the server farm while conveying the arranged quality of service (QoS). Specifically, in this, vitality productive figuring and propose (a) structural standards for vitality effective administration of clouds; (b) vitality proficient asset assignment approaches and planning calculations considering QoS desires and force use qualities of the gadgets; furthermore, (c) various open exploration challenges, tending to which can carry generous advantages to both asset suppliers and customers. This methodology was approved by leading a presentation assessment study utilizing the CloudSim toolbox. The outcomes exhibit that a cloud processing model has enormous potential as it offers huge cost-reserve funds and shows a high potential for the improvement of vitality proficiency under unique remaining task-at-hand situations.

Wang et al. (19) implemented an RA technique based on an auction in CC. For CC, an RA based on auction is presented first and discussed major issues in auction-based RA mechanism design. In CC, for the allocation of resources, it is a market-based mechanism. There should be a proper integration of CC characteristics and auction models for designing an auction-based RA mechanism.

Horri et al. (20) adopted a VM resource utilization history-based technique for implementing a QoS-aware VMs consolidation method. The CloudSim simulator is used for implementing, as well as evaluating, proposed algorithms. QoS metrics are enhanced, the consumption of energy is minimized as shown in the results of simulation and there is a trade-off between QoS and energy consumption in a cloud environment.

In a CC environment, for the allocation of resources, the various security concerns presented in the preceding section based on multiple perspectives and solutions for preventing co-resident attacks are classified as well as compared. Even with satisfying security requirements, their large-scale deployment is also considered. Efficiency will be decreased with additional control procedures. So there is a need to find a proper balance between security and performance.

14.3 PROPOSED METHODOLOGY

To reduce VM co-residency probability, Stackelberg game theory with threshold-based VM allocation strategy (SGT-TVMS) is proposed, and co-resident attacks cost are also increased using this strategy. For the VM migration-timing problem, SGTF is proposed, whereby the migration of the VM to various physical machines is decided by a cloud provider for minimizing the risk produced by collocated malicious VM.

Four thresholds are established in this work, and a co-residency-resistant VM allocation strategy is proposed, whereby threshold settings are changed. An allocation strategy can be changed dynamically for getting high-level security as well as better performance in allocation. The energy intake of resources, like storage, network, CPU and memory, has to minimized for enhancing energy efficiency. Limited resources are utilized effectively in the suggested strategies. For saving energy, idle resources are turned off and pushed into sleep mode.

14.3.1 Cloud System Model

In this model, the cloud is considered as a physical-machines set, and every machine is capable of hosting various amounts of VMs of various users. To assign VMs to physical machines, initially, a placement strategy is used by a cloud provider. The analysis is not affected by the placement strategy details, and the assumption is made that, control is not with any user. On the same physical machines, a set of victim VMs can be targeted by an adversary by collocating with them (21).

Through a game-theoretic framework, the interactions between adversary and cloud provider (defender) are studied here, and rewards are dependent on time. Specifically, to defend against collocation attacks, VMs are reassigned to various machines in the time selected by the defender's strategy. On the other side, the adversary selects an attack rate for launching more VMs for increasing its chances to have prolonged collocation with its victims. Three possible placement scenarios of the game are illustrated in Figure 14.1.

At time τ_a in plot (a), the VM of the attacker is collocated successfully with its target VM on same hypervisor before the migration of the target VM to some other ode at time τ_d. A successful collocation event is represented in this condition, whereby a leakage of information is produced.

Before placing the malicious VM, a migration of the target VM happens in plot (b), which indicates the avoidance of a collocation event.

At last, a no-migration policy is illustrated in plot (c), whereby there is a maximization in a collocation period.

14.3.2 Stackelberg Game Theory Steps

The formal presentation of the Stackelberg game model-based mitigation-timing problem is given in this section. For a security sensitive resource, there will be a fight between two players. A description of attack–defense scenarios are enabled using this model (22). Two classes of players are incorporated in this game who are competing against each other and termed defender and attacker. The defender is a player who protects resource, and the attacker is a player who tries to comprise resources. A detailed discussion about every game model element follows.

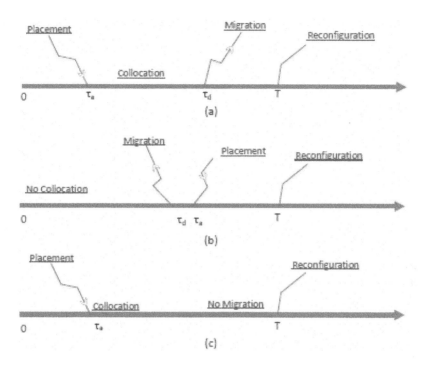

FIGURE 14.1   Three Possible Placement Scenarios for the Stackelberg Game Model.

A tuple $\Gamma(\rho, A, u)$ is used for defining the game, where

- ρ is a players' set. Here, $\rho = \{1, 2\}$, represents defender (player 1) and adversary (player 2)
- $A = A_d \times A_a$ is an action space of the defender and the adversary.
- $u = u_d, u_a$ is a reward function, $u : A \rightarrow R^2$.

14.3.2.1 Defender's Action Space

A reallocation period is controlled by the cloud provider, who is termed the system defender as stated from the timing factor investigation. Assume that a time constant as $\tau_d \in A_d$, whereby the defender migrates a running VM to a new physical node so that $A_d = [\tau_{min}, T]$. In this, the system parameter is represented as T, where the reset of a credential is allowed and the smallest reconfiguration time is represented as τ_{min}. At time T, assume a leakage model, whereby a reset of the credential happens, and from a side-channel attack, no benefit is given to the attacker.

To minimize the information leakage chances, the value of τ_d is optimized by the defender, and it avoids the system overloading with unnecessary migrations. Between stability and security, the trade-off optimization is a major goal of the defender. Specifically, the high security of the system is indicated by a small value for τ_d. This is because of the small co-residency times between any two VMs. But between physical nodes and VMs, frequent migration increases the overhead of the system.

The VM live migration overhead depends on the workload of VMs. The speed of network and the memory size of VM are major factors affecting overhead of VM migration as shown in this work. A more stable system is required, with high value of τ_d. On same node between VMs, the co-residency time is large, and it makes the system highly susceptible to data breaches through collocation attacks.

14.3.2.2 Attacker's Action Space

Here, an assumption is made that the system placement algorithms are not known to attacker, and so it tries to increase the request count submitted to provider of a cloud for increasing its co-residency chances. Assume $\lambda_a \epsilon A_a$ represents the request rate submitted to cloud, where the nonnegative attack rates interval is represented as $A_a = [\lambda_{min}, \lambda_{max}]$. At time t = 0, the game is assumed to start, and actual time is represented as τ_a, whereby the attacker successfully collocates with their targeted victim.

Hence, a nonnegative random variable is given by $\tau_a > 0$ with probability density function (pdf) $f_a(\tau_a :, \lambda_a)$, which is parametrized by λ_a. For every submitted job, the cost is paid by attacker, so it needs optimization over attack rate λ_a. So the trade-off of the attacker can be summarized as the probability of collocation with victim by attacker is less if λ_a is very small, λ_a is very small and any information stolen is taken before the migration of the VMs.

The successful collocation of the attacker is increased with a large value of λ_a at higher attack cost expenses. So, for yielding early collocation's highest probability, pdf f_a should be $f_a(\tau_a; \lambda_{a1})$ rather than $f_a(\tau_a; \lambda_{a2})$, when $\lambda_{a1} > \lambda_{a2}$.

14.3.2.3 Cost Function Calculation by Improved Support Vector Machine (ISVM)

In this work, the minimization of the defender's long-term discounted cost, which is defined as $\sum_{t=0}^{\infty} \alpha^t c(s_t)$, is considered as a major objective, where the discounted factor is given by $\alpha \in (0,1)$. Defenders are willing to minimize the cost in the current turn rather than minimizing in a future turn as interpreted from α. This is due to the fact that the next turn may be attacked by an attacker. The great patience of the defender is indicated using a high discount factor.

For a specified initial state s_0 and P, based on Markov chain, resource states involve with state space as V and the transition probability as P. So the discounted problem corresponds to the problem defender; there exists a coincidence between the transition probabilities and the defender's strategy. With an initial state s_0 = s, the long-term cost of defender can be rewritten as

$$C_P(s) = \sum_{t=0}^{\infty} c_p(s_t) \tag{14.1}$$

$$= C_p(s) + \alpha \sum_{s' \in N(s)} pss'E\left[\sum_{t=0}^{\infty} \alpha^t \left(s_{t+1}, s_{t+2}, \right) | s_1 = s'\right] \tag{14.2}$$

$$= C_p(s) + \alpha \sum_{s' \in N(s)} pss' C_P[s']. \tag{14.3}$$

14.3.2.3.1 Support Vector Machine (SVM)

Within the feature space, between the training sample's two classes is the best separating hyperplane, a plane having a maximum margin identified by using SVM by concentrating on cases positioned at class descriptors edges. Using this procedure, an optimum hyperplane can be computed, and it uses minimum training samples in an effective manner. With small training sets, it is possible to achieve high accuracy of classification (23).

Every example in mth class has a positive label, and all other examples, having negative labels, are used for training mth SVM. So, for specified data, where $(x_1, y_1), (x_2, y_2), \ldots (x_l, y_l), i = 1, 2, \ldots, l, where\, x_i \in R^l$ and $y_i \in \{1, 2, \ldots k\}$ is SVM class.

$$\phi(w, b, \xi) = \sum_{m=1}^{k} (w_m . w_m) + C\left(\sum_{i=1}^{l} \sum_{m \neq y_i}^{k} \xi_i^m\right) \tag{14.4}$$

In terms of $(w_{y_i} . \phi(x_j)) + b_{y_i} \geq (w_m . \phi(x_j)) + b_m + 2 - \xi_i^m$

$$\xi_i^m \geq 0, i = 1, \ldots \ldots l\, m \in \{1 \ldots \ldots, k\} \setminus y_i,$$

where, x is mapped onto a high dimensional space using a nonlinear function $\phi(x_i)$, weight vector is represented as w, bias is represented as b and the slack variable is given by ξ, a constant given by C and computed a priori. A type of quadratic programming problem corresponds to an optimum hyperplane search in Equation 14.4.

Between the attack data's two classes, a margin can be maximized using minimization of $\frac{1}{2}\sum_{m=1}^{k}(w_m . w_m)$. Training-errors count in nonlinearly separable data can be reduced using a penalty term $C\left(\sum_{i=1}^{l}\sum_{m \neq y_i}^{k}\xi_i^m\right)$. Balancing between training errors and regularization term is a basic concept of SVM. The k decisions functions are obtained by solving Equation 14.5.

$$f_k(x) = \sum_{i=1}^{l} \alpha_i^k y_i^k k(x, x_i) + b^k, \tag{14.5}$$

where a Lagrange multiplier and adaptive divergence weight kernel function are represented as kernel $k(x, x_i)$. In a data-dependent manner, an adaptive divergence weight kernel function $k(x, x_i)$ is modified for enhancing SVM classifier's accuracy in classification.

14.3.2.3.2 Adaptive Divergence Weight Kernel Function–
  Based Support Vector Machine (ADWK-SVM)

The ability to map nonlinearly separable data points implicitly into various dimensions is provided by kernels to support VMs, which makes them as linearly separable. The cost is involved in the mapping of data points as a high dimension. A large number of vectors, a large amount of memory and a longer time for computation are needed for high-dimensional data. In SVMs, there is no need of storing high-dimensional vectors explicitly.

Input is mapped onto a high dimension using it, and inner products will only be stored. Various mappings are provided by various kernel functions. There is no kernel's silver-bullet choice. There exist merits and demerits for every kernel. An available Gaussian kernel is modified in this work. When compared with the original Gaussian kernel, better performance is shown by this modified adaptive divergence weight kernel.

Through a nonlinear mapping ϕ, every input sample space R is mapped as a feature space F in nonlinear SVM. An embedding of S as F using mapping ϕ forms curve submanifold. In features space, S's samples are mapped using $\phi(x)$; small vector dx is mapped as

$$\phi(dx) = \nabla\phi, dx = \sum_i \frac{\partial}{\partial x^{(i)}}\phi(x)dx^{(i)}, \tag{14.6}$$

where $= \nabla\phi = \dfrac{\partial}{\partial x^{(i)}}\phi(x)$

The $\phi(dx)$'s squared length is written as

$$ds^2 = |\phi(dx)|^2 = \sum_{ij} g_{ij}(x)dx^{(i)}dx^{(j)}. \tag{14.7}$$

where $g_{ij}(x) = \dfrac{\partial}{\partial x^{(i)}}\phi(x).\dfrac{\partial}{\partial x^{(j)}}\phi(x).$

The summation over index α of ϕ is represented using dot. Positive-definite matrix $G(x) = g_{ij}(x))$ with $n \times n$ size is a Riemannian metric tensor, which is induced in S.

$$g_{ij}(x) = \frac{\partial}{\partial x^{(i)}}\frac{\partial}{\partial x^{(j)}} AK(x, x_j) \tag{14.8}$$

For enhancing SVM performance, distances (ds) or the margin between classes are increased. Around the boundary, the Riemannian metric tensor is increased, and it is reduced around other samples using Equation 14.8. Adaptive divergence weight kernel K can also be modified using Equation 14.8 so that $g_{ij}(x)$ is in large around the boundary.

The kernel modification according to Riemannian geometry structure is as follows: Assume the modifications of the kernel as

$$K(x,x_i) = p(x) px_i AK(x,x_i).$$ (14.9)

Equation 14.9 shows the kernel's conformal transformation by factor p(x). The adaptive divergence weight kernel function is used as a SVM kernel function:

$$AK(x,x_i) = exp\left(-(x-x_j)^2 / \sigma^2\right).$$ (14.10)

Here, kernel width is expressed as σ. It is shown that the respective Riemannian metric tensor is modified as

$$g_{ij}(x) = \frac{1}{\sigma^2}\delta_{ij}.$$ (14.11)

After the kernel Riemannian metric's modification, the tensor is expressed as

$$g_{ij}(x) = p_i(x)p_j(x) + p^2(x)g_{ij}(x).$$ (14.12)

Gaussian kernel's conformal transformation is used for ensuring a large value of p(x) around support vector (SV):

$$p_i(x) = \partial p(x)/\partial x_i.$$ (14.13)

For maximum p(x), the value of $p_i(x) = 0$.

For ensuring a large value if p(x) at the locations of support vector, it is designed as a data-dependent one:

$$p(x) = \sum_{i \in SV} \alpha_i exp\left(-(x-x_i)^2 / 2\tau^2\right),$$ (14.14)

where a free parameter is represented as α, and over all support vectors, summation is computed. If a is very near to support vectors, then values of $p(x)$ and $p_i(x)$ are high, and there will be an increase in $g_{ij}(x)$ around support vectors; if it is far away from support vectors, then their values will be very small. So there will be an enlargement of spatial resolution near the boundaries, and the SVM's ability in classification is strengthened. The following summarizes the proposed algorithm's procedure.

Step 1: With primary ADWK, SVM is trained for extracting SVs information; then, as per Equations 14.9 and 14.14, the kernel AK is modified.

Step 2: The SVM with modified kernel AK; the SVM is trained.

Step 3: Until achieving better performance, Steps 1 and 2 are applied iteratively.

14.3.2.4 Threshold-Based VM Allocation Strategy (TVMS)

A scenario is constructed for exploring the factors affecting VM co-residency probability as mentioned in Table 14.1. With respect to load balancing and security, the VM allocation strategy is analysed in the next section.

14.3.2.4.1 VM Allocation Strategy

In the cloud, users need to occupy a large number of hosts as required in the VM allocation strategy. On every host, the average VM count is expressed in Equation 14.15, and for evaluating load balancing, VM's standard deviation is expressed in Equation 14.16:

$$\underline{\mu} = \frac{\sum_{u \in U} |VMs(u)|}{M} \tag{14.15}$$

$$\sigma = \sqrt{\frac{\sum_{h \in H}(\sum_{u \in U} |\{v \mid v \in V, X_{v,u,h} = 1, h \in H\})^2}{M}}. \tag{14.16}$$

The previously mentioned standard deviation is defined by the same kind of hosts. There are various hosts types, so a separate computation of standard deviations of every host type is required. At last, these standard deviations' average value is computed. The minimization of VMs co-residency and analysis of factors affecting VMs co-residency are focused in this work. Based on Equations 14.17 and 14.18, S, M and $|Hosts(T)|$ are related to P.

There is a technique for guiding, as well as increasing, hosts' (M) size. A cloud platforms' input defines M, and the running hosts count is closely related with the consumption of power (24). So the minimization of M is not advisable. The VM count that can be generated by users is represented as S, as defined in Table 14.1.

TABLE 14.1
Variable Definitions

Name	Definition
V	VMs V = {$v_1, v_2, v_3 \ldots v_L$}
H	Hosts H = {$h_1, h_2, h_3 \ldots h_M$}
U	Users U = {$u_1, u_2, u_3 \ldots u_N$}
VMs(U)	The subset of VMs started by users U the maximum number of VMs is S.
$X_{h,u,r} = 1$	The VM v of user u is placed on the host h.
Hosts(U)	The subset of hosts occupied by users U. The total number of hosts in the cloud is M.
Users(V)	The subset of users who start VMs V
Co-Resident VMs (A,T)	The subset of VMs started by attackers A that are co-resident with VMs of target users T, A ⊆ U and T ⊆ U

So, for high-quality service, S is not required to be minimized. The minimization of $|Hosts(T)|$ is the only way to minimize the VMs' co-residency probability. The requirements of RA are entirely different from this.

$$P = 1\left(1 - \frac{|Hosts(T)|}{M}\right)^2 \tag{14.17}$$

$$P - 1 - \prod_{s=1}^{S}\left(1 - \frac{|Hosts(T)|}{M - |Hosts(s)|,}\right) \tag{14.18}$$

Malicious or legal users cannot be computed accurately using a cloud platform, but it is still possible to track malicious users. Large numbers of VMs are set up by a malicious user for occupying more hosts. So it is not possible to judge the attacker using VM count. There is a high chance of co-resident attacks with the created VM count via an increase in attackers.

There is a requirement to limit the host count occupied by an attacker, if a large number of VMs are created by attackers. If, throughout the entire allocation process, this is done with a small number of VMs, then the load balancing may not be satisfied by the cloud platform. So, in a VM allocation strategy, VM count plays an important role. The positioning of new VM is also associated using this with VMs count that user-initiated.

Consider VM Num of Host as a maximum VMs count on one host, and VMs and hosts type are having influence over it. Best load percentage and appropriate load rate are set by cloud providers in general for providing effective services. In an average case, this problem is considered as shown in Equation 14.19. User count on one host is represented as User Num of Host, and the average VM count started by one user is represented as VM Num of User:

$$VM\,Num\,of\,user = \frac{VM\,Num\,of\,Host \times Best\,load\,\%}{User\,Num\,of\,Host}. \tag{14.19}$$

As mention in the previous assumptions, a constant value is obtained in *VM Num of Host × Best load* %, and there will be an inverse-proportion relationship among *User Num of Host* and VM Num of One User. For minimizing VM co-residency, the *User Num of Host* should be minimized, and the *VM Num of User* should be increased as shown is the security analysis. The following factors should be considered in the secure VM allocation strategy design: VM count created by user, user count on one host and host count occupied by one user.

14.3.2.4.2 The Threshold of Strategy
While designing a VM allocation strategy, user count, host count and VM count should be considered through a security analysis. So, for balancing RA and security, four thresholds are set.

14.3.2.4.2.1 VM Num of Centralize   The $|Hosts(T)|$ minimization is the only way for minimizing VM co-residency. So honeypot hosts set to minimize $|Hosts(T)|$. Honeypot hosts are the co-resident attack detection platform. In real time, there will be monitoring of the VMs, which are positioned at honeypot hosts. If a certain threshold VM Num for Centralize is exceeded by a VM count started by one user, then the balance of VMs is placed on honeypot hosts. VM count on honeypot hosts is represented as VM Num in Honeypot, which is given in Equation 14.20:

$$VM\ Num\ in\ Honeypot = |\ VM(T) - VM\ Num\ for\ Centralize. \qquad (14.20)$$

The VM co-residency is reduced using honeypot hosts as fewer hosts are occupied by users. Different co-resident attack detection techniques can be implemented using honeypot hosts as it is a real-time monitoring platform. Malicious user range can be locked if attacks are detected. Else, after a particular span of time, without any attack on honeypots hosts, VMs can be migrated to hosts where users are already placed.

For honeypot deployment, running hosts are given priority from honeypot hosts. So, in practice, there won't be a dramatic increase in consumption of power and host count.

14.3.2.4.2.2 VM Num for Spread   The maximization of VM Num of User and the minimization of User Num of Host are considered in this section for initializing another threshold called VM Num for Spread according to VM count that a user has started. VMs are spread on more hosts by cloud platform if VM Num for Spread is greater than VM count. Else, VMs are spread over more hosts by the cloud platform, where this user's VMs are placed already. As in Equation 14.5, this problem is considered as an average case so that a user's VM Num value reaches $\frac{VM\ Num\ for\ Centralize}{VM\ Num\ for\ S;read}$. So, for centralization, VM Num can be adjusted by administrators and for better allocation resource or minimizing VMs co-residency, VM Num for Spread can be adjusted dynamically.

14.3.2.4.2.3 User Num in Host   From a server perspective, security risks are considered. In a cloud platform, co-resident attack risks are increased by allowing more users to deploy their VMs on one server. Malicious users can place their VMs on a target server if it has sufficient resources based on a previous VM allocation strategy. So an acceptable choice is to minimize the user count; for that User Num in Host, a threshold is set. It ensures that new users are not able to create a VM on this host.

14.3.2.4.2.4 Host Num for Spread   In a certain range, the places of VMs can be limited by hosts using the VM Num for Centralize and VM Num for Spread thresholds. There is a necessity to consider accurate host counts because of the impact shown by hosts' remaining resources and the allocation strategy's random factors that prevent VMs' location prediction by attackers using replaying experiments.

So, fourth threshold is introduced: Host Num for Spread, which provides an update regarding host count occupied by a user in real time. Like in VM Num for Spread, a user's new VM is deployed on a host if the host's count is more than Host Num for Spread, where VMs are placed before. But here, on various hosts, VMs are spread by hypervisor.

Four thresholds are set by analysing VM co-residency's influencing factors. They are User Num in Host, Host Num for Spread, VM Num for Centralize and VM Num for Spread. Based on these thresholds, a co-residency-resistant VM allocation strategy is proposed, which is a "first spread, later centralize and the more VMs create the more concentrate". As per Table 14.1, host count is defined as $|Hosts(A)|$, and it also indicates user VMs deployment; VM count started by user A in host m is defined as $|VMs(A)|$. Specifically, hosts and VMs' information are updated by hypervisor while user A is requesting to launch a VM.

For a better allocation of resources, a decentralized allocation strategy is selected by hypervisor, if threshold VM Num of Spread is not exceeded by VMs count started by user A as per this methodology. Else, VMs will be deployed by hypervisor in hosts where the VMs of a user are already deployed. Then VM Num for Spread's operation is similar to threshold Host Num for Spread.

If threshold VM Num is exceeded by $|VMs(A)|$, a honeypot mechanism will be initiated by hypervisor, and in honeypot hosts, VMs will be placed for preventing users potentially malicious behaviour to centralize. The VMs' deployment will be highly centralized with an increased count of VMs. User count in one host is limited to a certain value called User Num in Host, as mentioned in threshold User Num in Host. Using every weight, every host is weighted by scheduler; then weights are normalized and can be added up. At last, hosts are sorted by a scheduler, and the host with the largest weight value is selected.

14.4 RESULTS AND DISCUSSION

Extensive simulations are conducted for evaluating the proposed algorithm's performance. With respect to average attack duration, average migration time and detection accuracy, promising results are produced by the proposed algorithm as shown in the simulation results.

A detection accuracy performance metric comparison between available and proposed methodologies are shown in Figure 14.2. A detection accuracy of about 93.57% is produced in the proposed SGT-TVMS framework, which is greater compared to the a 91.87% detection accuracy produced by the SGTF, the 87.56% produced by the moving target defense (MTD) and the 81.24% by the Video Electronics Standards Association (VESA) Display Compression (VDC-M) methodology.

The migration time gap between 250 VMs and 200 VMs increases with an increase in physical machines (PMs) count as shown in Figure 14.3. This is because, with few PMs, for a specified 250 VM input, it is not possible to accept many VMs. There is no significant deviation between 200 VM input and 250 VM input. The cloud system can accept more VMs if there is an increase in PM hosts, which will increase the available resources. So the total time of migration can be increased.

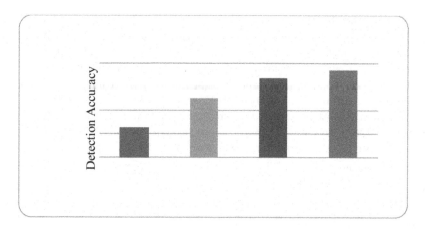

FIGURE 14.2   Performance Comparison of Detection Accuracy between the Proposed and Existing Techniques.

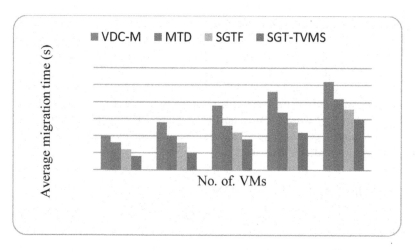

FIGURE 14.3   Performance Comparison of Total Completion Time against the Number of VMs.

An average attack duration metric comparison between available and proposed methodologies are shown in Figure 14.4. When compared with available VDC-M, MTD and SGTF approaches, attacks are detected effectively using the proposed SGT-TVMS framework as shown in that results.

14.5 CONCLUSION

Cloud providers require an RA strategy with less time for response while serving more users. To achieve user satisfaction in a cloud paradigm, it is required to have an effective RA methodology, and it also maximizes cloud service providers' profit.

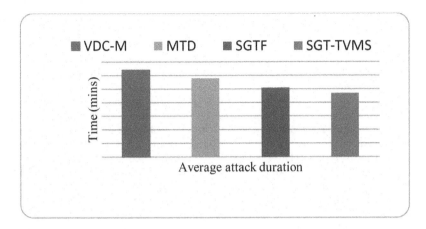

FIGURE 14.4 Average Attack Duration Comparison between the Proposed and Existing Techniques.

To reduce VM co-residency probability, a Stackelberg game theory with threshold-based VM allocation strategy (SGT-TVMS) is proposed in this work, and co-resident attacks cost are also increased using this strategy.

For VM migration-timing problem, SGTF is proposed, whereby the migration of VMs to various physical machines is decided by a cloud provider for minimizing the risk produced by collocated malicious VM. SGT-TVMS's probability was analysed in this work, and factors affecting the computation of co-resident VMs like user number, host number and VM number were computed.

Four thresholds were established, and a co-residency-resistant VM allocation strategy was proposed, whereby threshold settings are changed. An allocation strategy can be changed dynamically for getting a high level of security as well as better performance in its allocation. For a cloud provider, a better allocation of resources is provided using the proposed SGT-TVMS as shown in the results of the simulation, and when compared with available RA methodologies, better security against co-resident VMs is provided using the proposed technique.

14.6 ACKNOWLEDGMENTS

All the authors have equally contributed for writing this book chapter.

REFERENCES

1. Hameed, A. et al. (2016). A Survey and Taxonomy on Energy Efficient Resource Allocation Techniques for Cloud Computing Systems. *Computing*, 98(7), 751–774.
2. Beloglazov, A., and R. Buyya. (2012). Optimal Online Deterministic Algorithms and Adaptive Heuristics for Energy and Performance Efficient Dynamic Consolidation of Virtual Machines in Cloud Data Centers. *Concurrency and Computation: Practice and Experience*, 24(13), 1397–1420.

3. Kaur, K., and A. Pandey. (2013). ECO-Efficient Approaches to Cloud Computing: A Review. *International Journal of Advance Research in Computer Science and Software Engineering*, 3(3).

4. Beloglazov, A., and R. Buyya. (2010). Energy Efficient Resource Management in Virtualized Cloud Data Centers. *2010 10th IEEE/ACM International Conference on Cluster, Cloud and Grid Computing*, 826–831, May.

5. Lee, Y. C., and A. Y. Zomaya. (2012). Energy Efficient Utilization of Resources in Cloud Computing Systems. *The Journal of Supercomputing*, 60(2), 268–280.

6. Litvinov, V., and K. Matsueva. (2014). Resource-Efficient Allocation Heuristics for Management of Data Centers for Cloud Computing. *Proceedings of the National Aviation University*, 2, 113–118.

7. Akhter, N., and M. Othman. (2014). Energy Efficient Virtual Machine Provisioning in Cloud Data Centers. *2014 IEEE 2nd International Symposium on Telecommunication Technologies (ISTT)*, 330–334, November.

8. Panda, P. K., and S. Swagatika. (2012). Energy Consumption in Cloud Computing and Power Management. *International Journal of Advanced Research in Computer Science*, 3(2).

9. Amin, M. B. et al. (2015). Profiling-Based Energy-Aware Recommendation System for Cloud Platforms. In *Computer Science and Its Applications*. Berlin, Heidelberg: Springer, 851–859.

10. Chang, Y. C., S. L. Peng, R. S. Chang, and H. Hermanto. (2014). A Cloud Server Selection System—Recommendation, Modeling and Evaluation. In *International Conference on Internet of Vehicles*. Cham: Springer, 376–385, September.

11. Hasan, H. F. (2018). Enhanced Approach for Cloud Resource Allocation Using CPU Scheduling. *Iraqi Journal of Information Technology*, 8(2), 13–32.

12. Goel, N., I. P. Ncce, and I. S. Singh. (2017). Multiple Request Resource Allocation by Using Time-Shared Policy in Cloud Computing. *Journal of Network Communications and Emerging Technologies (JNCET)*, 7(7), www. jncet. org.

13. Bates, A. et al. (2012). Detecting Co-residency with Active Traffic Analysis Techniques. *Proceedings of the 2012 ACM Workshop on Cloud Computing Security Workshop*, 1–12, October.

14. Sundareswaran, S., and A. C. Squcciarini. (2013). Detecting Malicious Co-resident Virtual Machines Indulging in Load-Based Attacks. In *International Conference on Information and Communications Security*. Cham: Springer, 113–124, November.

15. Han, Y., T. Alpcan, J. Chan, C. Leckie, and B. I. Rubinstein. (2015). A Game Theoretical Approach to Defend Against Co-Resident Attacks in Cloud Computing: Preventing Co-Residence Using Semi-Supervised Learning. *IEEE Transactions on Information Forensics and Security*, 11(3), 556–570.

16. Zhang, Y., A. Juels, A. Oprea, and M. K. Reiter. (2011, May). Homealone: Co-Residency Detection in the Cloud via Side-Channel Analysis. *2011 IEEE Symposium on Security and Privacy*, 313–328.

17. Abazari, F., M. Analoui, and H. Takabi. (2017). Multi-Objective Response to Co-Resident Attacks in Cloud Environment. *International Journal of Information and Communication Technology Research*, 9(3), 25–36.

18. Beloglazov, A., J. Abawajy, and R. Buyya. (2012). Energy-Aware Resource Allocation Heuristics for Efficient Management of Data Centers for Cloud Computing. *Future Generation Computer Systems*, 28(5), 755–768.

19. Wang, H., H. Tianfield, and Q. Mair. (2014). Auction Based Resource Allocation in Cloud Computing. *Multiagent and Grid Systems*, 10(1), 51–66.

20. Horri, A., M. S. Mozafari, and G. Dastghaibyfard. (2014). Novel Resource Allocation Algorithms to Performance and Energy Efficiency in Cloud Computing. *The Journal of Supercomputing*, 69(3), 1445–1461.

21. Anwar, A. H., G. Atia, and M. Guirguis. (2019). A Game-Theoretic Framework for the Virtual Machines Migration Timing Problem. *IEEE Transactions on Cloud Computing*, https://par.nsf.gov/servlets/purl/10209203.
22. Wang, X., X. Chen, W. Wu, N. An, and L. Wang. (2015). Cooperative Application Execution in Mobile Cloud Computing: A Stackelberg Game Approach. *IEEE Communications Letters*, 20(5), 946–949.
23. Widodo, A., and B. S. Yang. (2007). Support Vector Machine in Machine Condition Monitoring and Fault Diagnosis. *Mechanical Systems and Signal Processing*, 21(6), 2560–2574.
24. Han, Y., J. Chan, T. Alpcan, and C. Leckie. (2014). Virtual Machine Allocation Policies Against Co-Resident Attacks in Cloud Computing. *2014 IEEE International Conference on Communications (ICC)*, 786–792, June.

Index